KEEPI COMPETITIVE U.S. MILITARY AIRCRAFT INDUSTRY ALOFT

Findings from an Analysis of the Industrial Base

JOHN **BIRKLER** | PAUL **BRACKEN**
GORDON T. **LEE** | MARK A. **LORELL**
SOUMEN **SAHA** | SHANE **TIERNEY**

Prepared for the Office of the Secretary of Defense

NATIONAL DEFENSE RESEARCH INSTITUTE

The research described in this report was prepared for the Office of the Secretary of Defense (OSD). The research was conducted within the RAND National Defense Research Institute, a federally funded research and development center sponsored by OSD, the Joint Staff, the Unified Combatant Commands, the Navy, the Marine Corps, the defense agencies, and the defense Intelligence Community under Contract W74V8H-06-C-0002.

Library of Congress Cataloging-in-Publication Data

Keeping a competitive U.S. military aircraft industry aloft : findings from an analysis of the industrial base / John Birkler ... [et al.].
    p. cm.
  ISBN 978-0-8330-5864-5 (pbk. : alk. paper)
  1. Airplanes, Military—Technological innovations—United States. 2. Competition—United States. 3. Military aeronautics equipment industry--United States. 4. Aircraft industry—Military aspects—United States. I. Birkler, J. L., 1944-

  TL685.3.K36 2011
  338.4'76237460973—dc23

                                                                        2011038271

The RAND Corporation is a nonprofit institution that helps improve policy and decisionmaking through research and analysis. RAND's publications do not necessarily reflect the opinions of its research clients and sponsors.

RAND® is a registered trademark.

Cover design by Pete Soriano
Cover vector drawings: The-Blueprints.com

Published 2011 by the RAND Corporation
1776 Main Street, P.O. Box 2138, Santa Monica, CA 90407-2138
1200 South Hayes Street, Arlington, VA 22202-5050
4570 Fifth Avenue, Suite 600, Pittsburgh, PA 15213-2665
RAND URL: http://www.rand.org/
To order RAND documents or to obtain additional information, contact
Distribution Services: Telephone: (310) 451-7002;
Fax: (310) 451-6915; Email: order@rand.org

Preface

The military aircraft industry in the United States is dominated by a handful of prime contractors. Whereas more than a dozen firms competed to develop and produce U.S. military aircraft during the first couple of decades after World War II, the industry consolidated dramatically in the 1980s, 1990s, and early 2000s such that today only three (or possibly four) domestic contractors develop, produce, and sustain complex fixed-wing military aircraft.[1] Moreover, only three major firms (Northrop Grumman, Raytheon, and BAE [British Aerospace Systems]) supply avionics, and just three others (General Electric, Rolls Royce, and Pratt & Whitney) produce large turbofan engines.

For at least two decades, policymakers have been expressing concerns that further consolidation could erode the competitive environment, which many believe is a fundamental driver of innovation in the military aircraft industry. Such concerns led to two congressionally mandated studies on the health and competitive prospects of the United States' fixed-wing military aircraft industrial base that RAND produced in 2003.[2]

---

[1] Boeing, Lockheed Martin, and Northrop Grumman are the only U.S. companies that produce modern manned aircraft for the military. General Atomics produces unmanned aircraft that perform surveillance and other tasks.

[2] John Birkler, Anthony G. Bower, Jeffrey A. Drezner, Gordon Lee, Mark Lorell, Giles Smith, Fred Timson, William P.G. Trimble, and Obaid Younossi, *Competition and Innovation in the U.S. Fixed-Wing Military Aircraft Industry*, Santa Monica, Calif.: RAND Corporation, MR-1656-OSD, 2003; Mark Lorell, *The U.S. Combat Aircraft Industry, 1909–2000*, Santa Monica, Calif.: RAND Corporation, MR-1696-OSD, 2003.

In the years since RAND published those studies, policymakers have continued to harbor concerns about the long-term health of the U.S. military aircraft industrial base, and in 2009 the Committee on Armed Services of the House of Representatives requested that RAND's 2003 analysis be updated.[3] This project, sponsored by the Under Secretary of Defense for Acquisition, Technology, and Logistics, responds to that request by updating the RAND study entitled *Competition and Innovation in the U.S. Fixed-Wing Military Aircraft Industry.* The project (1) reviewed that study's evaluations of the risks and costs of the United States having little or no competition among companies involved with designing, developing, and producing fixed-wing military aircraft and related systems; (2) examined changes in industrial-base structure and capabilities that have taken hold since that analysis was performed; and (3) assessed how these and future changes will affect the industrial base.

This monograph should be of interest to policymakers concerned with military aircraft design, development, and production and with aerospace industrial base issues. It was sponsored by the Office of the Secretary of Defense and conducted within the Acquisition and Technology Policy Center of the RAND National Defense Research Institute, a federally funded research and development center sponsored by the Office of the Secretary of Defense, the Joint Staff, the Unified Combatant Commands, the Navy, the Marine Corps, the defense agencies, and the Defense Intelligence Community.

A companion volume will update *The U.S. Combat Aircraft Industry, 1909–2000,* the other study that RAND produced in 2003.

For more information on the RAND Acquisition and Technology Policy Center, see http://www.rand.org/nsrd/ndri/centers/atp.html or contact the director (contact information is provided on the web page).

---

[3] See U.S. House of Representatives, National Defense Authorization Act for Fiscal Year 2010: Report of the Committee on Armed Services House of Representatives on HR-2647 Together with Additional and Supplemental Views, Washington, D.C.: U.S. Government Printing Office, June 18, 2009, p. 380.

Contents

Figures

Tables

Summary

A handful of prime contractors dominate the United States' military aircraft industry today. Whereas during the first several decades after World War II, more than a dozen firms competed to develop and produce U.S. military aircraft, now only three domestic contractors develop, produce, and sustain complex fixed-wing military manned aircraft. One major firm supplies unmanned aircraft, three major firms supply avionics, and three contractors produce large turbofan engines.

For at least two decades, policymakers have expressed concerns that further consolidation could erode the competitive environment for military aircraft and degrade the industry's abilities to develop, manufacture, and support innovative designs. In 2001, at the request of the U.S. Senate, the Department of Defense (DoD) asked RAND's National Defense Research Institute to study the implications of having little or no competition in the fixed-wing military aircraft industry. RAND performed that evaluation and published its results in 2003.[4]

Policymakers' concerns have persisted in the years since publication of that study, and in 2009 the Committee on Armed Services of the House of Representatives requested that RAND's 2003 analysis be updated. This project responds to that request. Carried out for the Under Secretary of Defense for Acquisition, Technology, and Logistics, the project reviewed RAND's earlier evaluation of the risks and costs of the United States' having little or no competition among companies involved with designing, developing, and producing fixed-wing military aircraft and related systems; examined changes in industrial-base

---

[4] See Birkler et al., 2003, and Lorell, 2003.

structure and capabilities that have taken hold since that analysis was performed; and determined how these and future changes will affect the industrial base.

To conduct the study, we interviewed and collected data from three major prime fixed-wing aircraft contractors—the Boeing Company, Lockheed Martin Corporation, and Northrop Grumman Corporation—and from General Atomics, the main manufacturer of unmanned aerial systems (UAS). We also communicated with EADS (European Aeronautic Defence and Space Company N.V.) North America. In addition, we held discussions with and collected data from various DoD offices and numerous other organizations.[5] The data we collected allowed us to update the database that we used in the 2003 study,[6] which we then used to populate models from which we projected the impact that combinations of new aircraft programs that are not in DoD's current procurement pipeline would have on the industrial base.

Current Status of the Fixed-Wing Military Aircraft Industrial Base

Three major prime manufacturers—Boeing, Lockheed Martin, and Northrop Grumman—dominate the domestic fixed-wing industry

---

[5] We interviewed individuals at Boeing, Lockheed Martin, Northrop Grumman, EADS North America, and U.S. government offices. We used proprietary data from all those companies except EADS North America. We also obtained other data from the Aerospace Industries Association, previous RAND work, the Commission on the Future of the United States Aerospace Industry, the National Science Foundation, the Institute for Defense Analyses, the Office of the Deputy Assistant Secretary of Defense for Industrial Policy, the Office of the Under Secretary of Defense (Comptroller), service acquisition commands and laboratories, and company annual reports.

[6] The database drew from program budget exhibits—R-1 documentation for RDT&E data and P-1 documentation for procurement data. It also included Selected Acquisition Reports, Budget Item Justification exhibits, and other budget and planning documents. However, one difference from our 2003 report was that we were not able to break out by funding between prime and major subcontractors. We were able to do that in 2003, but because these numbers change frequently, such refinements were not part of the current study's database.

in the United States today. Another company, General Atomics, has arisen in the past decade as the main prime in the UAS field.

However, the industry continues to evolve. Thirty years ago, companies focused on manufacturing airframes and platforms; 20 years ago they concentrated on providing integrated systems. Today, however, they largely provide system integration capabilities. Primes now outsource much of what they once did in house. They do, however, maintain sufficient core skills to oversee and support their second-tier vendors. At the same time, the aerospace industry appears to be morphing toward commercial enterprise models that rely on networks of agile, smaller teams that have autonomy, budgets, and delayered authority structures and processes. This means that with primes increasingly focused on integrating complex systems, significant innovations are now expected to occur in second-tier firms as well as at the prime contractor level.

Assessment Criterion

We stayed close to the intent of Congress in choosing the criterion to gauge the adequacy of the U.S. military fixed-wing aircraft industrial base. We used the legislative language "that the United States must ensure, among other things, that more than one aircraft company can design, engineer, produce and support military aircraft in the future."[7] We interpreted that language to mean that the U.S. industrial base would be adequate if it was able to sustain at least two full-service prime contractors, each possessing approximately equal shares of *both* research, development, test, and evaluation (RDT&E) funding and procurement funding.[8]

---

[7] U.S. House of Representatives, 2009, p. 380.

[8] This equal split is supported by the Herfindahl-Hirschman Index, a commonly accepted measure of market concentration. The index takes into account the relative market shares and distribution of the firms in a market and approaches zero when a market has a large number of firms possessing relatively equal shares of the market. The index increases as the number of firms in the market decreases and as the disparity in the market shares between those firms increases. The index is at minimum when firms have equal shares of the market. For further explanation, see U.S. Department of Justice, "The Herfindahl-Hirschman Index," n.d.

Current Research and Procurement Funding Is High

In 2010, the industrial base operated in an environment in which annual funding for RDT&E was at a 30-year high, hitting $13.45 billion, and funding for procurement was at $32.23 billion, twice the level of a decade earlier.[9]

Figures S.1 and S.2[10] display RDT&E and procurement outlays (in billions of fiscal year [FY] 2011 dollars) for programs funded from FY 1980 through FY 2010.[11]

Figure S.1
Fixed-Wing Military Aircraft RTD&E Funding, FY 1980–2010

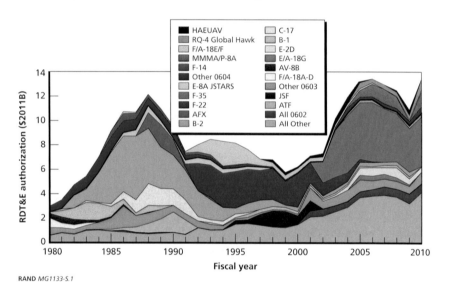

RAND *MG1133-S.1*

---

[9] Amounts are in fiscal year (FY) 2011 dollars. Two elements make up procurement funding: aircraft production ($22.75 billion in 2010) and aircraft modifications ($9.51 billion in 2010). Much of the increase in RDT&E has been due to the F-35 program, which is developing three versions of the Joint Strike Fighter.

[10] For these and subsequent RDT&E and procurement funding figures, we used data from program budget exhibits—R-1 documentation for RDT&E data and P-1 documentation for procurement data. We also used Selected Acquisition Reports, Budget Item Justification exhibits, and other budget documents.

[11] The "All Other" category in Figures S.2 and S.3 contains a multitude of smaller programs.

Figure S.2
Fixed-Wing Military Aircraft Procurement Funding, FY 1980–2010

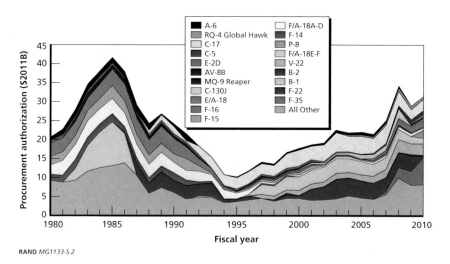

RAND *MG1133-S.2*

The four primes received only a portion of the funding depicted in Figure S.1 and Figure S.2. Moreover, while the distribution of RDT&E funding has been uneven, procurement funding has been more evenly balanced in recent years (See Figures S.3 and S.4).[12]

When these figures are viewed in the light of the assessment criterion discussed above, current programs of record (as of 2010) do not adequately appear to sustain two or more primes, each receiving

----

[12] Readers should note that the RDT&E funding data that we received did not break out the funds that primes allocate to subcontracts. Thus, Lockheed Martin's subcontracts to Northrop Grumman on the F-35, for example, were not broken out in our data set. As a result, the data we display may overrepresent Lockheed Martin's RDT&E share and underrepresent Northrop Grumman's share. Similarly, in terms of procurement, the amounts shown are for the contracted prime only and do not separately identify subcontracted work, such as Northrop Grumman's participation in the F/A-18 and F-35 programs.

Figure S.3
RDT&E Funding for Prime Contractors, FY 2000–2010

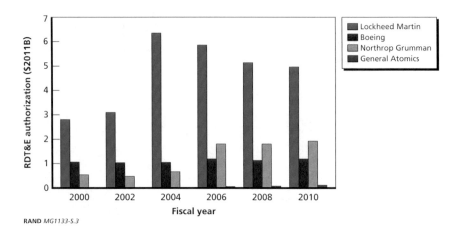

RAND *MG1133-S.3*

Figure S.4
Procurement Funding for Prime Contractors, FY 2000–2010

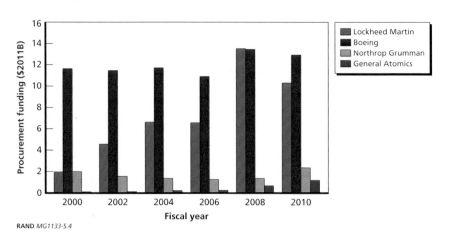

RAND *MG1133-S.4*

approximately equal shares[13] of both RDT&E funding and procurement funding.[14]

What If DoD Were to Pursue Additional Programs?

We identified six new programs that DoD might consider pursuing: the T-X trainer, the KC-X tanker, the unmanned carrier-launched surveillance and strike aircraft (UCLASS), the F-22 fighter sold as foreign military sales (termed F-22 FMS), a next-generation bomber, and a sixth-generation fighter.[15]

Using the 2011 Future Years Defense Plan as our funding base case, we modeled the degree to which these six programs might foster innovative, competitive conditions in the future. We found that if DoD were to award three new programs—T-X, KC-X, and UCLASS—to the contractor most needing the work to maintain its viable competitive status as a designer and producer of military aircraft (in this case, Boeing), the industry still would cease to be competitive after 2015.[16] That also would be the outcome if, in addition to those three programs, DoD were to pursue foreign military sales of the F-22.

---

[13] Although the congressional language motivated us to split RDT&E and procurement funding equally between two primes, it is not clear exactly what the shares should be. Depending on the circumstances, unequal divisions of funding—say 60:40, 70:30, or even 80:20—may be sufficient to sustain multiple primes for a period of time. Additionally, there also may be circumstances where funding could be split among three primes, either equally or unequally. However, if sustained over the long term, such unequal divisions may put lesser-funded primes at a disadvantage.

[14] Readers who compare the current report and the 2003 document should be aware of a fundamental difference between the two studies. The previous study used Selected Acquisition Reports (SARs), contract data, and available contractor reporting data to estimate the allocation of total obligation authority among prime contractors when two (or more) were involved in specific programs (e.g., F/A-18, F-22, JSF/F-35, etc.). Because these numbers change frequently, such estimates were not done for the current study. Consequently, charts depicting contractor funding levels/shares are not comparable between the two studies.

[15] In this monograph, we use the acronym *FMS* to denote both foreign military sales and other export sales to non-U.S. customers.

[16] Boeing generates $64 billion in total revenues, of which only 22 percent comes from unclassified military aircraft contracts.

However, by involving two primes equally in performing RDT&E and procurement on a next-generation bomber, DoD could sustain two firms through 2020 with RDT&E funding and through 2025 with procurement funding (see Figures S.5–S.8). Note that the KC-X does not appear in Figures S.5 and S.6 but does appear in Figures S.7 and S.8. The reason is that funding for the new tanker's RDT&E is already in the authorized budget, whereas funding for procurement has yet to be decided.

Adding a sixth-generation fighter to the previous industrial base cases would have funding impacts similar to the next-generation bomber. Assuming the program is shared between Lockheed Martin and Northrop Grumman, the RDT&E base would be sustained through 2025 with two primes having almost equal shares in the latter years. In terms of procurement, the fighter would not have much of an impact until the middle of the next decade, with three primes having almost equal shares (see Figures S.9 and S.10).

Figure S.5
RDT&E Funding: Base Case Plus T-X, F-22 FMS, UCLASS, and Next-Generation Bomber Programs, FY 2000–2025

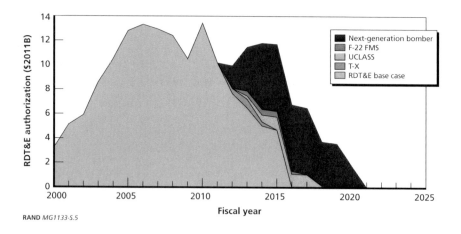

RAND MG1133-S.5

Figure S.6
Potential Prime Contractor Shares of RDT&E Funding: Base Case Plus T-X,
F-22 FMS, UCLASS, and Next-Generation Bomber Programs, FY 2010–2025

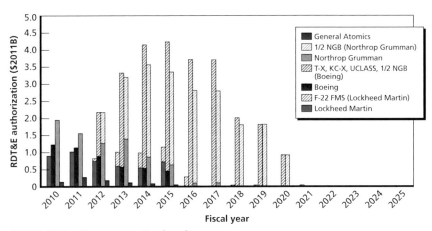

NOTES: NGB = Next-generation bomber.

RAND *MG1133-S.6*

Figure S.7
Procurement Funding: Base Case Plus T-X, F-22 FMS, UCLASS, and Next-
Generation Bomber Programs, FY 2000–2025

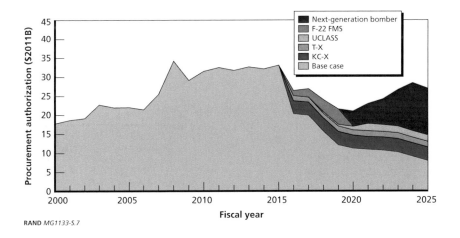

RAND *MG1133-S.7*

Figure S.8
Potential Prime Contractor Shares of Procurement Funding: Base Case Plus T-X, F-22 FMS, UCLASS, and Next-Generation Bomber Programs, FY 2010–2025

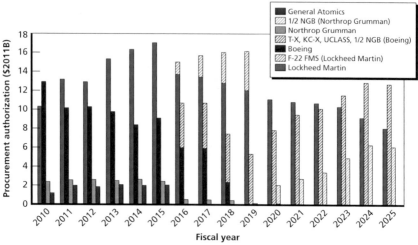

NOTE: NGB = Next-generation bomber.
RAND *MG1133-S.8*

Figure S.9
Potential Prime Contractor Shares of RDT&E Funding: Base Case Plus T-X, F-22 FMS, UCLASS, Next-Generation Bomber, and Sixth-Generation Fighter, FY 2010–2025

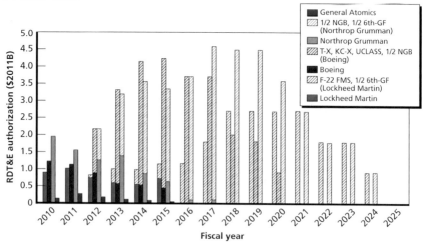

NOTES: NGB = Next-generation bomber. 6th-GF = 6th-generation fighter.
RAND *MG1133-S.9*

Figure S.10
**Potential Prime Contractor Shares of Procurement Funding: Base Case Plus
T-X, F-22 FMS, UCLASS, Next-Generation Bomber and Sixth-Generation
Fighter, FY 2010–2025**

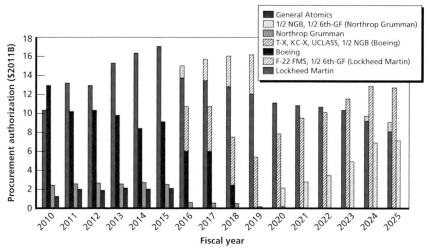

NOTES: NGB = Next-generation bomber. 6th-GF = 6th-generation fighter.
RAND MG1133-S.10

Which Program Combinations Would Best Sustain Competition?

Our evaluations suggest that small programs likely will not sustain
the industrial base, in terms of either RDT&E or procurement.[17] The
T-X, KC-X, and UCLASS programs would, in combination, sustain
only Boeing as a viable competitor in the fixed-wing military market
if it were the winner. Selling the F-22 to foreign militaries would
boost Lockheed Martin's fortunes for only four years or so (FY 2016–
2019).

To keep two primes healthy and competitive through 2025,
a next-generation bomber program, or a program of similar size, is
needed. This could sustain Boeing and Northrop Grumman if each

---

[17] However, as noted in RAND's 2003 study, a series of relatively small demonstration programs could sustain the advanced design teams that are precursors to new major systems.

company were to share 50 percent of funding for RDT&E and procurement. After 2025, an additional program on the scale of a sixth-
generation fighter would need to be pursued. Table S.1 displays how
the primes fare under each of these strategies, with cells in gray and
yellow denoting combinations that would significantly or modestly
sustain the primes, respectively.[18]

Table S.1 suggests that it may be possible to maintain a competitive and innovative fixed-wing military aircraft industrial base into the
immediate future, even with a reduced number of prime contractors
and with new program starts during a period of growing pressure on
the federal budget. This is mainly because of three new trends that have

Table S.1
Program Combinations That Would Sustain Primes in 2011–2025 Period and Post-2025 Period

Time period	Strategy	Boeing	Lockheed Martin	Northrop Grumman
2011–2025	T-X + KC-X + UCLASS	▓		
	+ F-22 foreign military sales	▓	░	
	+ next-generation bomber	▓	░	▓
Post-2025	+ sixth-generation fighter	▓	░	▓

NOTE: Gray = significantly sustained; yellow = modestly sustained.

---

[18] While it was outside the scope of this study, we recognize that attention must be paid to
the distribution of activities among contractors. Simply dividing funding equally between
two contractors does not guarantee that all the activities that companies need to remain
robust across the design-support spectrum will be nurtured. Under conditions in which few
programs are under way and their schedules do not overlap, the opportunities for such nur-
turing may not easily arise, inasmuch as long gaps may separate opportunities presented to
specific skill groups in companies. With more programs, keeping an advanced design group
or a field support activity going becomes easier because groups of personnel can be moved
from one project to the next.

emerged over the past ten years: the dramatic upsurge in RDT&E and procurement funding following the attacks of September 11, 2001; the large increase in the development and procurement of UAS, accompanied by the entry of new contractors and regeneration of traditional firms; and the continuing movement toward greater competitive outsourcing of research, development, and production tasks to lower-tier contractors, both foreign and domestic.

The future composition and capabilities of the military aircraft industry depend largely on the amount of business that the industry receives from DoD and how that business is distributed among development of technology, development of new designs, and production of completed designs. In Table S.1, we show the case in which firms that most need the work will win the award.

Competition may not produce the outcome displayed, however, and the industry may become further concentrated. Directed shares may be necessary to sustain multiple primes into the foreseeable future. Unless very purposeful and structured program decisions are made soon, the congressional objective—maintaining two or more companies capable of designing, engineering, producing, and supporting military aircraft—may not be achieved.

Acknowledgments

This research was greatly facilitated by the assistance of many individuals throughout the U.S. Department of Defense and the major aerospace firms—Boeing, Lockheed Martin, Northrop Grumman, and EADS North America. Their names and contributions would fill several pages. The data they provided and the insights they shared were critical to our quantitative analysis and to our interpretations and conclusions described in this report.

We particularly wish to thank Dawn Vehmeier, Manufacturing and Industrial Base Policy, Under Secretary of Defense (Acquisition, Technology and Logistics), for her support and assistance throughout the study.

We also wish to thank former RAND colleagues Kevin Brancato and Christine Osowski for their earlier contributions and to Fred Timson and Mike Thirtle for their thoughtful reviews that occasioned many changes and improved the clarity of the report.

Abbreviations

BAE	British Aerospace Systems
CAD	computer-aided design
DMS	diminishing manufacturing sources
DoD	Department of Defense
DUSC(IP)	Deputy Under Secretary of Defense for Industrial Policy
EADS	European Aeronautic Defence and Space Company N.V.
EMD	engineering and manufacturing development
FFRDC	federally funded research and development center
FMS	foreign military sales
FY	fiscal year
FYDP	Future Years Defense Program
GT	group technology
IPT	integrated product team
ISR	intelligence, surveillance, and reconnaissance
IT	information technology
JPG	Joint Programming Guidance
NDAA	National Defense Authorization Act
NGB	next-generation bomber
P&G	Procter and Gamble

PE	private equity
POM	Program Objectives Memorandum
PPBE	planning, programming, budgeting and execution
PPBS	Planning, Programming, and Budgeting System
R&D	research and development
RDT&E	research, development, test, and evaluation
S&E	science and engineering
SAR	Selected Acquisition Report
SECDEF	Secretary of Defense
SPG	Strategic Planning Guidance
TOA	total obligation authority
UAS	unmanned aerial systems
UAV	unmanned air vehicle
UCAV	unmanned combat air vehicle
UCLASS	unmanned carrier-launched surveillance and strike aircraft
VC	venture capital

CHAPTER ONE

Introduction

A handful of prime contractors dominate the U.S. fixed-wing military aircraft industry. In the first several decades after World War II, more than a dozen firms competed to develop and produce U.S. military aircraft. But since then, the industry has consolidated dramatically. Today, only three domestic contractors (possibly four) develop, produce, and sustain complex fixed-wing military aircraft.[1] Only three major firms (Northrop Grumman, Raytheon, and BAE [British Aerospace Systems]) supply avionics, and just three others (General Electric; Rolls Royce, the owner of Allison Engine Co.; and Pratt & Whitney) produce large turbofan engines.

For at least two decades, policymakers have been expressing concerns that further consolidation could erode the competitive environment, which many believe is a fundamental driver of innovation in the military aircraft industry. The issue crystallized in the fall of 2001 when the Department of Defense (DoD) chose Lockheed Martin to be the prime contractor to develop and manufacture the Joint Strike Fighter, known as the F-35. The F-35 is the only new major combat aircraft program that the United States currently is pursuing.

Even before DoD chose Lockheed Martin as the F-35 prime contractor, senior DoD officials and members of Congress had begun to voice concerns about the effect of that contract award on the ability of all three U.S. prime contractors to remain as active designers and pro-

---

[1] Boeing, Lockheed Martin, and Northrop Grumman are the only U.S. companies that produce modern manned aircraft for the military. General Atomics produces unmanned aircraft that perform surveillance and other tasks.

ducers of military aircraft and on their long-term ability to operate in competitive and innovative ways. In December 2001, the U.S. Senate requested (in the DoD Appropriations Act of 2002) that DoD prepare a comprehensive analysis of and report on the risks to innovation and cost of limited or no competition in contracting for military aircraft and related weapon systems for the Department of Defense. RAND performed that evaluation and published its results in 2003.[2]

Adding to these concerns are more recent predictions of changes in the military aircraft procurement landscape. In March 2009, for example, the Deputy Under Secretary of Defense for Industrial Policy (DUSD[IP]) predicted that "[o]ver the next five to ten years, most current military aircraft production programs will end, precipitating the need for a new round of consolidation in order to reduce infrastructure costs." Furthermore, "[t]he reduction in RDT&E funding does not bode well for companies without long term production programs."[3] (In DoD parlance, RDT&E funding stands for monies devoted to research, development, test, and evaluation.)

Responding to these recent concerns, the Committee on Armed Services of the House of Representatives, reporting in June 2009 on the National Defense Authorization Act (NDAA) for Fiscal Year (FY) 2010, expressed its desire that

- prime contractors and suppliers remain competitive and innovative and be cost-efficient
- more than one aircraft company be able to design, engineer, produce, and support military aircraft in the future.[4]

---

[2] John Birkler, Anthony G. Bower, Jeffrey A. Drezner, Gordon Lee, Mark Lorell, Giles Smith, Fred Timson, William P.G. Trimble, and Obaid Younossi, *Competition and Innovation in the U.S. Fixed-Wing Military Aircraft Industry*, Santa Monica, Calif.: RAND Corporation, MR-1656-OSD, 2003; Mark Lorell, *The U.S. Combat Aircraft Industry, 1909–2000*, Santa Monica, Calif.: RAND Corporation, MR-1696-OSD, 2003.

[3] U.S. Department of Defense, Office of the Under Secretary of Defense for Acquisition, Technology, and Logistics Industrial Policy, *Annual Industrial Capabilities Report to Congress*, March 2009.

[4] U.S. House of Representatives, National Defense Authorization Act for Fiscal Year 2010, Report of the Committee on Armed Services House of Representatives on HR-2647

The committee directed the Secretary of Defense to commission a study by a federally funded research and development center (FFRDC) to update RAND's 2003 analysis, "particularly in light of DoD programmatic decisions made in the last seven years and the recent DUSD(IP) assessment."

In a separate but related mandate, the final NDAA for FY 2010 directed the Secretary of Defense also to report on the impact on the industrial base of developing an exportable version of the F-22A:[5]

> (c) ADDITIONAL REPORT REQUIRED.—The Secretary of Defense shall enter into an agreement with a federally funded research and development center to submit, not later than 180 days after the date of the enactment of this Act, to the committees identified in subsection (a), through the Secretary of Defense, a report on the impact of foreign military sales of the F-22A fighter aircraft on the United States aerospace and aviation industry, and the advantages and disadvantages of such sales for sustaining that industry.

Research Objective and Approach

This monograph documents our response to both congressional directives. In the body of the monograph, we update RAND's 2003 analysis of the U.S. fixed-wing military aircraft industrial base, which evaluated the risks and costs of the United States having little or no competition among companies involved with designing, developing, and producing fixed-wing military aircraft and related systems; examined changes in industrial-base structure and capabilities that have taken hold since that analysis was performed; and determined how these and future changes will affect the industrial base. In Appendix A we evaluate the impact of foreign military sales of the F-22.

Together with Additional and Supplemental Views, Washington, D.C.: U.S. Government Printing Office, June 18, 2009, p. 380.

[5] *National Defense Authorization Act for Fiscal Year 2010*, Public Law 111-84, October 28, 2009, Section 1250.

In our analysis, we have attempted to stay close to congressional concerns as expressed in legislation. Thus, we focused on maintaining the present competitive structure and capabilities of the current prime contractors. We confined our analysis to fixed-wing aircraft, drawing on unclassified information.

What Does the Industrial Base Entail?

In this monograph, we use the term *industrial base* broadly. The U.S. fixed-wing military aircraft industrial base includes the entire nation's capabilities of designing, engineering, producing, and sustaining fixed-wing military aircraft. These capabilities combine a vast array of scientific and engineering knowledge with business discipline and are partly the result of decades of military R&D and procurement funding.

The industrial base in which these capabilities are embedded is composed of a variety of organizations, from government research labs, test centers, and repair depots to the complex hierarchy of private firms that own or manage the facilities, equipment and tools, processes, designs, and patents and that employ the skilled labor force with the requisite experience and human capital. Taking action to "sustain" or "enhance" the industrial base means consciously choosing which capabilities and which organizations should receive sustained or new funding.

Prime Contractors

The fixed-wing military aircraft industrial base can be broken down into a complex hierarchy of firms.[6] At the top are the prime contractors for manned fixed-wing aircraft: Boeing, Lockheed Martin, and Northrop Grumman. Over the past ten years they have been joined in the mid-size unmanned fixed-wing aircraft arena by General Atomics.[7]

---

[6]   RAND's previous research described the long history of the consolidation of the prime contractors up to 2002. Firm organizational structure has been relatively constant since that period, and is not emphasized here.

[7]   All the manned fixed-wing military aircraft primes are pursuing unmanned systems. As of FY 2011, only Northrop Grumman has a program with significant RDT&E and procurement funding: the Global Hawk.

Contracts for RDT&E and production are provided to prime contractors, which are ultimately responsible for developing and producing the aircraft. Primes can choose to use in-house capacity and capabilities or can partner or subcontract with other firms for all or part of the aircraft and its subsystems, testing, support equipment, and training. It is important to note that no prime contractor has ever been, or ever can be, completely "full-service." Often, prime contractors must partner with companies in the same tier and/or with subcontractors in lower tiers to produce aircraft subsystems or components.

First-Tier and Second-Tier Suppliers

Below the prime contractor level are first-tier and second-tier suppliers of parts and subsystems. An aircraft can be broken down into many elements—airframe structure, vehicle systems, mission systems, and engines. Some of these have historically been designed and produced by the prime contractors; others have been outsourced to other firms. In the past, subcontracted parts and subsystems would generally be shipped to a prime contractor's facility for integration, final assembly, checkout, and testing. But, as discussed in Chapter Two, that practice has begun to change, and the primes are overseeing these activities, some of which might be conducted by other parties.

Assessment Criterion

We stayed close to the congressional intent in choosing the criterion by which to gauge the adequacy of the U.S. military fixed-wing aircraft industrial base. We used the legislative language "that the United States must ensure, among other things, that more than one aircraft company can design, engineer, produce and support military aircraft in the future."[8] For the purposes of this analysis, we interpreted that language to mean that the U.S. industrial base would be adequate if it were able to sustain at least two full-service prime contractors, each

---

[8] U.S. House of Representatives, 2009, p. 380.

possessing approximately equal shares of *both* RDT&E funding and procurement funding.[9]

Research Tasks

We translated Congress' directives into five research tasks:

- *Task 1: Describe the current status of the fixed-wing military aircraft industrial base.* This task involved determining the capabilities of the U.S. military aircraft industrial base and identifying ongoing programs, their levels of activity, and their duration, as well as trends and likely changes that will affect the industry's ability to provide innovative and cost-effective systems.

- *Task 2: Evaluate ways to encourage innovation in light of recent experience.* The linkage between competition and innovation is not well defined, and neither competition nor innovation can be directly measured in analytically satisfying ways. In this task, we sought to better understand factors affecting competition and innovation so that defense policymakers can provide a posture that ensures a continued high level of innovation in the future. We especially sought to understand competitive pressures as a stimulus to technological innovation and to investigate innovative and non-innovative industrial sectors, identifying those industry attributes and characteristics that contribute to successful innovations and sustain enduring competition. We evaluated whether recent historical experience would cause us to change or modify the paradigm of innovation used in the previous RAND study.

- *Task 3: Assess prospects for innovation and competition in the military aircraft industry.*[10] In this task, we examined how the level

---

[9] While we recognize that primes also engage in tasks, activities, and other elements that they need to share, we used funding as a quantifiable measure on which to base our criterion.

[10] The aircraft industrial base has a unique economics problem in the sense that it is not a competitive market by definition of many suppliers and many demanders. On the contrary, it tends to be specialized by weapon system type; in today's world, there may only be a single supplier of a specific system at the prime level. Likewise, the government could be considered to be a monopsony in terms of its demand function: Commercial air carriers or businesses do not demand fighter aircraft for their operations; the military is the only customer for that

and composition of demand for military aircraft might change over the next decade, and how such changes would affect the structure, competitiveness, and overall levels of capability of the industry. A critical issue we examined is the minimum level and content of business required to sustain a firm so that it is capable of functioning successfully as a prime contractor for a military aircraft program.

- *Task 4: Evaluate the effects of F-22 foreign military sales (FMS) on the industrial base.*[11] We examined the effects on the industrial base of selling the F-22 to foreign militaries. How would export of an F-22 FMS version affect F-22 prime contractors? How would it affect the demand for competitive systems, and what would the net impact be on the industrial base?
- *Task 5: Identify policy options open to DoD.* In this task, we assessed policy options available to DoD to guide the evolution of the industry and ensure maintenance of critical abilities and characteristics.

Research Methodology

To perform these tasks, we followed a four-track methodology:

- Review current literature on the aircraft industrial base.
- Update the database used in our 2003 report, which contains information and statistics on programs, funding, and schedules for the range of activities that constitute RDT&E and procurement.
- Explore the literature on innovation and how the approaches that industrial organizations are taking with respect to innovation have evolved in the past decade.

---

type of product. For a fuller discussion of competition in the military aerospace arena, see Birkler et al., 2003; and Lorell, 2003. See also John Birkler, Mark V. Arena, Irv Blickstein, Jeffrey A. Drezner, Susan M. Gates, Meilinda Huang, Robert Murphy, Charles Nemfakos, and Susan K. Woodward, *From Marginal Adjustments to Meaningful Change: Rethinking Weapon System Acquisition*, Santa Monica, Calif.: RAND Corporation, MG-1020-OSD, 2010.

[11] In this monograph, we use the acronym *FMS* to connote both foreign military sales and other export sales to non-U.S. customers.

- Collect information from and conduct formal and informal interviews with executives from the three major prime fixed-wing aircraft contractors: Boeing, Lockheed Martin, and Northrop Grumman. As part of this track, we met with executives of General Atomics, a leading designer and manufacturer of unmanned aerial systems (UAS), and EADS (European Aeronautic Defence and Space Company N.V.) North America. We also held discussions with various DoD offices and numerous other organizations, which provided substantial supporting information and insights.

The entities we met with are outlined in Table 1.1.

Organization of the Monograph

Following this Introduction, Chapter Two discusses the current status of the U.S. fixed-wing military aircraft industrial base. That is followed in Chapter Three by a discussion of ways to encourage innovation in light of recent developments in the U.S. military fixed-wing industrial base. Chapter Four goes on to detail prospects for innovation and competition in the industrial base. Finally, Chapter Five provides the findings of the analysis, evaluates policy options open to DoD, and offers concluding comments. These chapters address Tasks 1–3 and Task 5.

Three appendixes follow these chapters. Appendix A addresses Task 4 by discussing the implications for U.S. industry of selling the F-22 fighter to non-U.S. customers. Appendix B compares the RDT&E and procurement budget projections that we made in our 2003 report with how those budgets actually fared between 2003 and 2010. And Appendix C shows the total value of the new RDT&E and procurement funding that the alternative new programs discussed in Chapter Five would engender. Whereas Chapter Four shows this funding divided among various primes, Appendix C aggregates it as a total without shares apportioned to primes, and Appendix D displays planned military aircraft procurement inventories through FY 2021. A bibliography completes the monograph.

Table 1.1
Firms and Organizations RAND Contacted

Interviews	Boeing
	Lockheed Martin
	Northrop Grumman
	General Atomics
	EADS North America
	Government
Proprietary Data	Boeing
	Lockheed Martin
	Northrop Grumman
	Government
Other	Aerospace Industries Association
	Previous RAND work
	Aerospace Commission
	National Science Foundation
	Institute for Defense Analyses
	Office of the Secretary of Defense/Industrial Policy (OSD/IP) annual reports
	OSD comptroller
	Service acquisition commands and labs
	Company annual reports

The Current Status of the Fixed-Wing Military Aircraft Industrial Base in the United States

This chapter provides an overview of the fixed-wing military aircraft industrial base in the United States as it existed in 2010. It discusses changes in the industry's structure, in the nature of the programs, and in DoD funding that have taken place since RAND's 2003 report on the industrial base.

To gain insight into the military aircraft industry, we looked at

- its organization and structure
- its total DoD funding, measured in total obligation authority (TOA) program[1] funding, measured by RDT&E funding and by procurement funding for modifications and production
- its business practices.

Aerospace Industry Organization and Structure

On the surface, the number of major companies in the industrial base for fixed-wing military aircraft is unchanged from the number that made up the industrial base that RAND studied in 2003.[2] Lockheed Martin, Boeing, and Northrop Grumman remain the three prime

---

[1] *Total obligation authority* pertains to the funds corresponding to the total budget authority across DoD or some specified part of it in a given year.

[2] For the purposes of this study, the U.S. military fixed-wing aircraft industrial base includes all people, firms, tools, facilities, and knowledge required to innovate, design, develop, produce, and sustain the most advanced aircraft systems.

contractors capable of developing advanced aircraft systems. Figure 2.1 shows the consolidations occurring over the past half century that have resulted in this three-prime environment.

Figure 2.1 also shows the appearance of one new player in the field, General Atomics, which arose in the past decade as the dominant prime in the UAS field.

As Figure 2.1 implies, the industry continues to change. Thirty years ago, prime contractors focused on manufacturing airframes and platforms; 20 years ago, they concentrated on providing integrated systems. Today, however, they largely provide system capabilities. They have moved away from maintaining complete aircraft design and manufacturing capacities and have transferred much of that work to second-tier companies and to non-U.S. firms. But they do maintain sufficient core skills that are necessary to oversee and support their second-tier vendors.

This change has entailed a major divestment among the leading prime contractors of many areas of traditional development and manufacturing work, which have migrated to first- or second-tier contractors, both in the United States and overseas. Aerospace primes now routinely hold competitions for major platform subsystems and components among first- or second-tier contractors. In many cases, the competing subcontractors are provided only with performance require-

Figure 2.1
U.S. Military Aircraft Industry Prime Contractors, 1960–2010

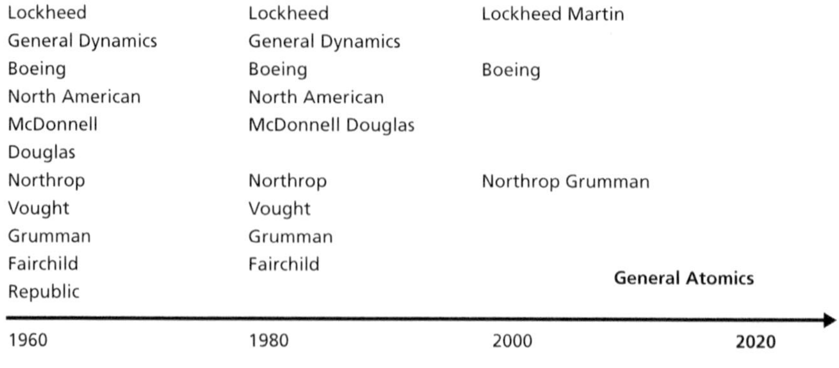

Lockheed	Lockheed	Lockheed Martin	
General Dynamics	General Dynamics		
Boeing	Boeing	Boeing	
North American	North American		
McDonnell	McDonnell Douglas		
Douglas			
Northrop	Northrop	Northrop Grumman	
Vought	Vought		
Grumman	Grumman		
Fairchild	Fairchild		
Republic		**General Atomics**	
1960	**1980**	**2000**	**2020**

ments and form, fit, and function requirements. Thus, the winning subcontractor must conduct its own design, development, and manufacturing for the component or subsystem.[3]

This trend is taking place in both the commercial and military aerospace worlds. Thus, in the case of the Lockheed Martin F-35, Lockheed's share in dollar terms of the production program is less than 20 percent. During research and development (R&D), many subcontractors were also responsible for conducting the development work for their own subsystem or components.

Since 2003, no major consolidation of primes or of primes with first-tier suppliers has occurred. However, this stability masks considerable underlying evolution: a shift in funding toward programs that meet current needs, tremendous uncertainty in future force structure priorities on the part of both government and contractor organizations, and contractor emphasis on spreading risk and responsibility throughout the supply chain.

Recent Trends in Total DoD Funding for Military Aircraft

In 2010, the industrial base—which, in addition to these three major primes and the main UAS prime, also includes a myriad of other smaller firms—was operating in an environment in which RDT&E funding was at a 30-year high and procurement funding had doubled since 2000. Table 2.1 contrasts the number of primes in 2010 with those in 2003.

---

[3] Despite these trends, readers should bear in mind that innovations in the aerospace arena are occurring across the system, section, subsystem, and equipment spectrum. Traditionally, many innovations related to military aircraft have taken place at the prime contractor level, as exemplified by stealth, high-angle of attack controllable flight, and supersonic speeds. Innovations by non-prime companies have enabled many system-level performance characteristics, such as engines (e.g., supersonic and supercruise) and avionics (e.g., active electronically scanned array radar). But the importance of funding and task division between contractors in circumstances when few programs are under way can have a significant effect on the ability to maintain functions that are the locus of system-level innovation.

Table 2.1
U.S. Primes Working on DoD Military Aircraft Programs, 2003 and 2010

Year	Number of Primes	Number of Production Programs	RDT&E Funding (billions of FY 2011 dollars)	Procurement Funding (billions of FY 2011 dollars)	
				Production	Modifications
2003	4	23	10.26	18.36	5.16
2010	4	26	13.45	22.75	9.51

Table 2.1 shows that between 2003 and 2010, funding for RDT&E, production, and modifications climbed 31 percent, 24 percent, and 84 percent, respectively.

Figures 2.2–2.5 display the programs receiving that funding and the level of funding (in billions of FY 2010 dollars) between 1980 and

Figure 2.2
Fixed-Wing Military Aircraft RTD&E Funding, FY 1980–2010, Sand Display

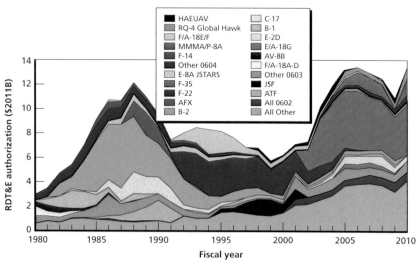

Figure 2.3
Fixed-Wing Military Aircraft RTD&E Funding, FY 1980–2010, Line Display

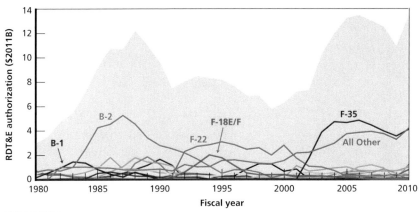

RAND *MG1133-2.3*

Figure 2.4
Fixed-Wing Military Aircraft Procurement Funding, FY 1980–2010, Sand Display

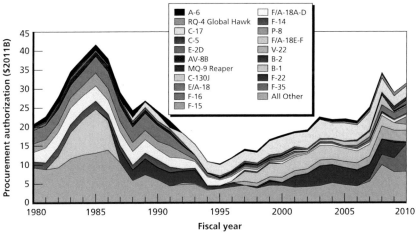

RAND *MG1133-2.4*

Figure 2.5
Fixed-Wing Military Aircraft Procurement Funding, FY 1980–2010, Line Display

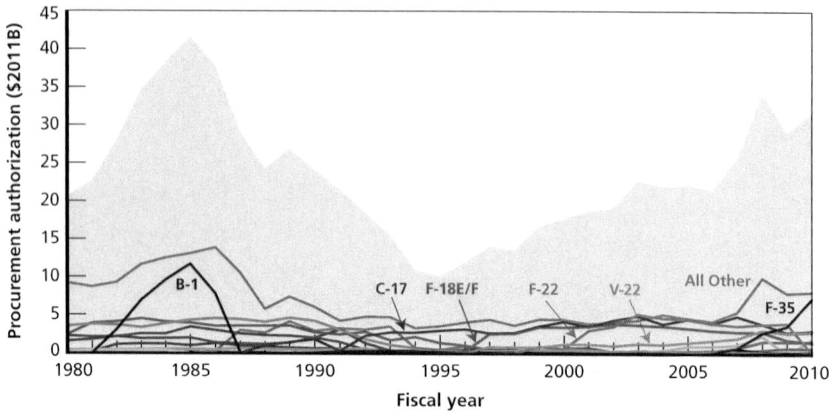

RAND *MG1133-2.5*

2010.[4] Figures 2.2 and 2.3 display programs' RDT&E funding, the former depicting the data cumulatively stacked in a sand display, the latter showing data for each program as an individual line. Figures 2.4 and 2.5 display programs' procurement funding in the same fashion. The gray shading in Figures 2.3 and 2.5 denotes the cumulative total from the figure immediately preceding.

Readers will note that in both the RDT&E and procurement data displays, the category "All Other" is quite significant. This category contains a multitude of smaller programs, as displayed in Table 2.2.

As the figures show, several lines continue to produce manned military aircraft for DoD: The F-35 is in low-rate initial production, the E-2D has just begun production, and the F/A-18 and C-130J production lines remain active. However, the F-22 is nearing the end of production before shutdown, and the C-17 appears to be on its last

---

[4] For these and subsequent RDT&E and procurement funding figures throughout the monograph, we used data from program budget exhibits—R-1 documentation for RDT&E data and P-1 documentation for procurement data. We also used Selected Acquisition Reports, Budget Item Justification exhibits, and other budget documents.

Table 2.2
Programs in "All Other" Category, RDT&E and Procurement, 1980–2010

	Procurement	
RDT&E	**Production**	**Sustainment**
Various "squadron program" entries for aircraft depicted in Figure 2.4 and Figure 2.5 and for B-52, KC-135, EC-135, F-4, F-111, A-10, F-117A, U-2	F-5, F-8 A-7, A-10, A-12 U-2 A/H/W/LC-130	B-52 FB-111 F-5, F-106, F-111, F-117, F-35 A-3, A-4, A-7, A-10, A-37
U-2	P-3	U-2
JSTARS funding with PE of 02xx rather than 06xx	RQ-7 MQ-8	V-22 C-1, C-2, C-5, C-9, C-12, C-18,
Assorted research into unmanned flight falling under no direct program	S-3 C-2, C-5, C-9, C-20, C-27, C-29, C-32,	C-20, C-21, C-25, C-29, C-32, C-37, C-40, C-130, C-135, C-137, C-141
Engine and avionics testing	C-37, C-40 KC-10	KC-10
	E-6	E-3, E-4, E-8
	Trainers	OV-10
	Business jets	MQ-1
	Civil air patrol, post-production support, electronic countermeasures, initial spares	Trainers Modification installation, special operations support, civil air reserve, aircraft subsystems, war consumables

NOTE: JSTARS = Joint Surveillance and Target Attack Radar System.

production lot for DoD, except for a few lots for export. The F-15 and F-16 production lines remain open for export customers.[5]

Since RAND's 2003 study, medium-to-large-size unmanned aircraft have become more prominent. After 300 units were procured, 2009 was the last year of procurement of the MQ-1 Predator. The MQ-9 Reaper is now being procured in large quantities (up to 48 per year). The RQ-4 Global Hawk is being procured in small quantities (four to five a year). The Navy is currently investing in technologies for carrier landing aircraft.

---

[5]   Readers should note that funds for export sales are not captured in the Future Years Defense Program (FYDP) data that we used to generate our procurement charts.

It should be noted that modifications make up a significant part of procurement spending. As shown in Figure 2.6, modifications accounted for about 30 percent of procurement funding in 2010, up from about 20 percent in 2003.

Examining RDT&E and procurement funding in light of overall TOA funding provides a slightly different picture. Plotted as a percentage of DoD's TOA as shown in Figure 2.7, procurement funding for fixed-wing military programs, while declining from the mid-1980s to the mid-1990s, rebounded to essentially the same level in 2010 as it was in 1980. However, the share of TOA represented by the programs' RDT&E funding nearly tripled during the same period.

But, as shown in Figure 2.8, over the same time period fixed-wing military aircraft accounted for 10–20 percent of all of the funding that DoD devoted to RDT&E and 18 to nearly 30 percent of the funding that it devoted to procurement.

Table 2.3 provides a detailed snapshot of the elements that made up the FY 2011 RDT&E program. Programs whose funding is less than $200 million account for roughly one-third of all RDT&E dollars.

Figure 2.6
Procurement Funding Shares: Modifications and Production Aircraft,
FY 2003–2010

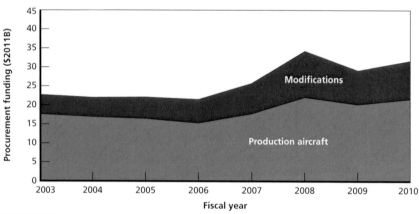

Figure 2.7
Percentage of DoD TOA Funding for Fixed-Wing Military Aircraft RTD&E
and Procurement, FY 1980–2010

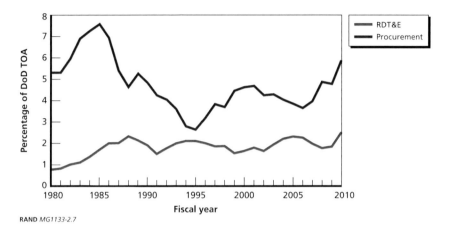

RAND *MG1133-2.7*

One thing to note from Figure 2.9 and Figure 2.10: If we use the assessment criterion that we discussed in Chapter One, current programs of record (as of 2010) do not adequately appear to sustain two or more primes, each receiving approximately equal shares[6] of both RDT&E funding and procurement funding.[7]

---

[6] It is not clear exactly what the shares should be. Depending on the circumstances, unequal divisions of funding—say 60:40, 70:30 or even 80:20—may be sufficient to sustain multiple primes for a period of time. Additionally, there also may be circumstances where funding could be split among three primes, either equally or unequally. However, if sustained over the long term such unequal divisions may put lesser-funded primes at a disadvantage.

[7] Readers who compare the current report and the 2003 document should be aware of a fundamental difference between the two studies. The previous study used Selected Acquisition Reports (SARs), contract data, and available contractor reporting data to estimate the allocation of total obligation authority between prime contractors when two (or more) were involved in specific programs (e.g., F/A-18, F-22, JSF/F-35, etc.). Because these numbers change frequently, such estimates were not done for the current study. Consequently, charts depicting contractor funding levels/shares are not comparable between the two studies.

Figure 2.8
Percentage of DoD RTD&E and Procurement TOA Represented by
Fixed-Wing Military Aircraft, FY 1980–2010

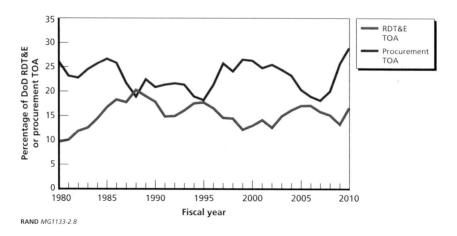

RAND *MG1133-2.8*

Figure 2.9
RDT&E Funding for Prime Contractors, FY 2000–2010

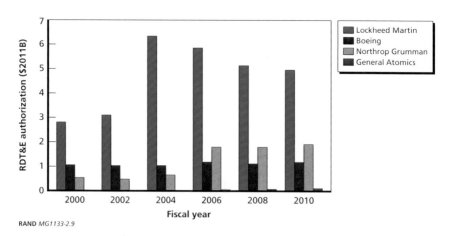

RAND *MG1133-2.9*

Figure 2.10
Procurement Funding for Prime Contractors, FY 2000–2010

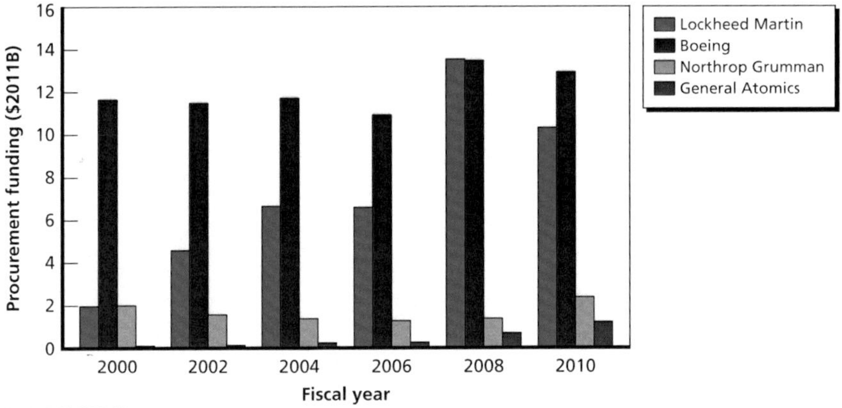

RAND *MG1133-2.10*

Table 2.3
FY 2011 RDT&E Funding Breakout
(millions of FY 2011 dollars)

Program Element Title	2011 Funding
Joint Strike Fighter SDD	2,073
Joint Strike Fighter (JSF)	1,948
Multi-mission Maritime Aircraft	1,171
F-22 Squadrons	569
RQ-4 UAV	439
Aerospace Propulsion and Power Technology	414
B-2 Squadrons	407
Advanced Hawkeye	361
Defense Research Sciences	328
F-15E Squadrons	320
Global Hawk UAV	317
Unmanned Combat Air Vehicle (UCAV)	305

Table 2.3—Continued

Program Element Title	2011 Funding
Advanced Aerospace Systems	258
Tactical Unmanned Aerial Vehicles	211
Joint Surveillance and Target Attack Radar System	186
Electronic Warfare Development	185
Materials	179
Airborne Warning and Control System	176
Airborne SIGINT Enterprise	167
C-17 Aircraft	162
Airborne Reconnaissance Systems	145
B-1B Squadrons	143
F-16 Squadrons	143
Aircraft Engine Component Improvement Program	140
Aerospace Vehicle Tech	139
Aerospace Sensors	136
Aviation Improvements	135
F/A-18 Squadrons	121
Air/Ocean Tactical Applications	118
Next Generation Jammer	117
Aviation Advanced Technology	112
C-130 Airlift Squadron	109
B-52 Squadrons	102
MQ-9 UAV	97
Aerospace Technology Development/ Demonstration	88
C-5 Airlift Squadrons	85

Table 2.3—Continued

Program Element Title	2011 Funding
FCS Reconnaissance (UAV) Platforms	75
Special Operations Aviation Systems Advanced	72
Advanced Materials for Weapon Systems	68
Fighter Tactical Data Link	67
E-2 Squadrons	63
Department of Defense Unmanned Aircraft	61
EA-18	57
Airborne Reconnaissance Systems EP-3	55
Advanced Aerospace Sensors	53
Manned Reconnaissance Systems	52
Automated Air-to-Air Refueling	43
MQ-1 Predator A UAV	37
KC-10s	36
Electronic Combat Technology	32
C-130J Technology	30
Aviation Survivability	27
Large Aircraft IR Countermeasures (LAIRCM)	27
E-4B National Airborne Operations Center (NAOC)	26
MQ-8 UAV	26
AV-8B Aircraft—Engine Development	21
HC/MC-130 Recap RDT&E	21
CV-22	20
CSAR-X RDT&E	15
Multi-Platform Electronic Warfare Equipment	15

Table 2.3—Continued

Program Element Title	2011 Funding
Next Generation Aerial Refueling Aircraft	15
Air Crew Systems Development	13
Special Operations CV-22 Development	13
A-10 Squadrons	12
EP-3E Replacement	12
Aerospace Electronic Attack–EA-6B	11
KC-135	10
Manned Destructive Suppression	10
Tactical Airborne Reconnaissance	10
Deployable Joint Command and Control	9
Joint Cargo Aircraft	9
Aviation Safety Technologies–2010 Collision Avoidance	8
MC103J SOF Tanker Recapitalization	6
Operational Support Airlift	5
Aviation Engineering Analysis	4
P-3 Modernization Program	4
RQ-11 UAV	1
RQ-7 UAV	1
Unmanned Vehicles	1

NOTES on next page.

Table 2.3—Continued

NOTES: EMD = engineering and manufacturing development;
IR = infrared; FCS = Future Combat Systems; SIGINT = signals
intelligence; SOF = special operations forces. The prime contractors
received only a portion of the funding depicted in Figures 2.2 and
2.3. Note that when the funding is assigned to prime contractors,
as depicted in Figures 2.9 and 2.10, their RDT&E funding is unevenly
distributed. Procurement funding displays a more even balance in
recent years between two of the primes. The RDT&E funding data
that we received did not break out the funds that primes allocate to
subcontracts. For example, Northrop Grumman's subcontract value
on the F-35 is included in the Lockheed Martin total in our data set.
As a result, the data we display may overrepresent Lockheed Martin's
RDT&E share and underrepresent Northrop Grumman's share.

Military Aircraft Industry Business Practices

In this section, we look at the U.S. fixed-wing military aircraft industrial base from a global perspective, paying particular attention to
changes over the last decade. We first take a quick snapshot of market
size and composition, then examine changes in firm structure and look
at how the business model of U.S. and European prime contractors
in the fixed-wing aircraft industry has changed. The discussion is not
exclusively about the military aircraft industry, however, since trends
in the commercial aircraft industry—and in the overall aerospace
industry—have been driving changes on the military aircraft market.

Global Aerospace Market

The global aerospace market is large, growing, and highly competitive.
Its revenue was $675 billion in 2008, with annual growth of 5.4 per-
cent from 2004 to 2008. In 2008, defense accounted for 69.5 percent
of this market; civil aerospace, the remaining 30.5 percent. The American market (which includes North and South America) accounts for
51.9 percent; the European Union (EU), 27.3 percent; and the Asia-
Pacific area, 18.5 percent. No single firm dominates the overall market:
Boeing leads with a 10.3 percent market share, followed by EADS

(7.6 percent), Lockheed Martin (6.5 percent), and Northrop Grumman (4.8 percent).[8]

U.S. aircraft manufacturers are heavily integrated in the global marketplace and depend heavily on international markets for sales and sourcing.[9] The U.S. aerospace industry sold more than $95 billion in aerospace vehicles and equipment to overseas customers while importing over $37 billion in aerospace products from abroad. However, these top-level figures do not indicate the nature, size, stability, ownership, or organization of the firms that are in the market, or how firms have changed over time.

Global Firm Dynamics

In the 1980s, U.S. prime manufacturers led the world market, in the design and production of both commercial and military aircraft. On the commercial side, Boeing, Martin-Marietta, McDonnell, Douglas Aircraft, General Dynamics, and Lockheed supplied over 80 percent of the world's needs.[10] At that time, the Soviet Union produced some commercial aircraft, primarily used by Soviet bloc countries; a few other aircraft were manufactured by Dassault, British Aerospace, Fokker, Embraer, and Bombardier. On the military side, McDonnell Douglas, General Dynamics, Lockheed, Boeing, Northrop, Grumman, and Rockwell fulfilled most of the U.S. and allied military requirements for fighters, bombers, and cargo aircraft. The Russians produced fighters, bombers, and other aircraft to meet their requirements and those of some other countries.

In the early 1990s, several major aerospace giants merged. In the United States, the result was an oligopolistic triumvirate of Boeing, Lockheed Martin, and Northrop Grumman as the prime companies for both commercial and military aircraft. European industry saw a

---

[8] Datamonitor USA, *Global Aerospace & Defense Industry Profile*, New York: Datamonitor Publication 0199-1002, December 2006.

[9] Michaela D. Platzer, *U.S. Aerospace Manufacturing: Industry Overview and Prospects*, Washington, D.C.: Congressional Research Service, CRS Report R40967, December 3, 2009.

[10] Estimated.

similar transformation under the banner of Airbus Industries for commercial aircraft and EADS for military aircraft (EADS includes firms from Britain, France, Netherlands, Germany, Spain, and Italy).[11] The worldwide commercial market is now dominated by a duopoly of Boeing and Airbus Industries. Although Boeing has not lost production quantities over the past 20 years, it has lost a significant share of the world commercial market to Airbus. According to company figures, Boeing booked 530 orders in 2010, nearly the same as it had booked in 1990. Over the same period, Airbus orders climbed to 574 from 404.[12]

Since the 1990s, the U.S. fixed-wing military aircraft industry has rapidly globalized. European entrants—BAE and Alenia, but not Boeing—have substantial parts of the Joint Strike Fighter program, which is the single largest military aircraft production program over the next decade. Furthermore, BAE Systems and Alenia have teamed with the EADS consortium to produce the Eurofighter and other military aircraft. EADS also has joined those companies in producing commercial aircraft under Airbus Industries.

The emerging global business model for the U.S. aerospace industry is amply demonstrated by two of the latest aircraft programs: the F-35 Joint Strike Fighter and the Boeing 787 Dreamliner. Figure 2.11 depicts the companies in the eight partner countries that were members of the JSF's global supply chain team in 2006.[13] Figure 2.12 shows the Boeing 787 Dreamliner's global supply chain.

---

[11] Until 2001, Airbus was a marketing consortium established under French law as a "Groupe d'Intérêt Economique." The four shareholders—Aerospatiale-Matra (37.9 percent), British Aerospace (20 percent), Construcciones Aeronauticas (4.2 percent), and Daimler Aerospace (37.9 percent)—performed dual roles as owners and industrial contractors. Most major decisions required unanimous approval of the shareholders. Airbus was obliged to distribute production work among its shareholders according to political as well as economic considerations. Then, Airbus was reorganized into a single fully integrated limited company. The objective was to streamline operations across national boundaries, reduce costs, and speed production.

[12] See Boeing and Airbus websites for order histories.

[13] The number of partner countries involved in the supply team could possibly grow to nine. See CAPT John Martins, *Joint Strike Fighter Program Update*, slide presentation, n.d.

Figure 2.11
F-35 Global Supply Sources, 2006

SOURCE: CAPT John Martins, *Joint Strike Fighter Program Update*, F-35 Lightening II
Program Office, slide presentation, n.d.
RAND *MG1133-2.11*

Tables 2.4 and 2.5 provide more details about the 787 Dreamliner's domestic suppliers and non-U.S. suppliers, respectively. The tables suggest that the 787 Dreamliner supply chain is driven primarily by affordability considerations. By way of contrast, the F-35 procurement, though proclaimed as "best value," is slightly constrained by U.S. commitments to provide meaningful work to its partner countries in exchange for aircraft to be procured by each country.[14]

---

[14] As traditionally understood, the offsets are not a consideration for the F-35. However, partner countries are evaluating the value of in-country work while determining the number of aircraft to be procured. The UK's BAE Systems is one partner in the F-35 program, manu-

Figure 2.12
Boeing 787 Dreamliner Global Supply Chain

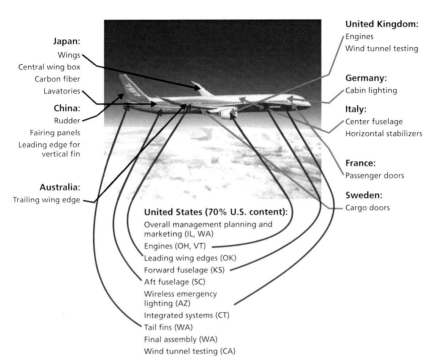

Japan:
Wings
Central wing box
Carbon fiber
Lavatories

China:
Rudder
Fairing panels
Leading edge for
vertical fin

Australia:
Trailing wing edge

United Kingdom:
Engines
Wind tunnel testing

Germany:
Cabin lighting

Italy:
Center fuselage
Horizontal stabilizers

France:
Passenger doors

Sweden:
Cargo doors

United States (70% U.S. content):
Overall management planning and
marketing (IL, WA)
Engines (OH, VT)
Leading wing edges (OK)
Forward fuselage (KS)
Aft fuselage (SC)
Wireless emergency
lighting (AZ)
Integrated systems (CT)
Tail fins (WA)
Final assembly (WA)
Wind tunnel testing (CA)

SOURCE: Dick K. Nanto, *Globalized Supply Chains and U.S. Policy*, 2010.
RAND *MG1133-2.12*

Vertical Integration and Globalization

In the mid-1980s, the major aerospace companies were vertically integrated. In addition to final assembly and checkout, each company had

---

facturing the aft fuselage and the fuel system (design and manufacturing responsibility); Alenia Aeronautica of Italy will start a final assembly and checkout (FACO) line for the final assembly of about 700 aircraft. Turkish Aerospace Industries (TAI) is scheduled to manufacture 400 of the center fuselages for Northrop Grumman. These are in addition to about 80 percent of composite parts/assemblies and many vehicle systems end items that are being planned to be procured from Australia, Canada, Denmark, Israel, Italy, The Netherlands, Norway, and Turkey. However, all software and mission system end items are being kept strictly within the United States.

Table 2.4
Domestic Suppliers for 787 Dreamliner

Company/Business Unit	Main Location	787 Work Statement
Boeing Commercial Airplanes (announced November and December 2003)	Washington	Airplane development, integration, final assembly, program leadership
Boeing Charleston (announced as Vought Aircraft Industries, November 2003)	South Carolina	Aft fuselage
Spirit Aerosystems, Inc. (announced as Boeing–Wichita, November 2003; April 2004)	Kansas, Oklahoma	Fixed and movable leading edges, flight deck, part of forward fuselage, engine pylons
Hamilton Sundstrand (announced February 2004, March 2004, July 2004, September 2004)	Connecticut	Auxiliary power unit, environmental control system, remote power distribution units, electrical power generating and start system, primary power distribution, nitrogen generation, ram air turbine emergency power system, electric motor hydraulic pump subsystem
Rockwell Collins (announced February 2004, June 2004)	Iowa	Displays, communications/ surveillance systems, pilot control system
Honeywell (announced February 2004, July 2004, December 2004)	Arizona	Navigation, maintenance/crew information systems, flight control electronics, exterior lighting
Goodrich (announced March 2004, April 2004, June 2004, November 2004, December 2004)	North Carolina	Fuel quantity indicating system, nacelles, proximity sensing system, electric brakes, exterior lighting, cargo handling system
Boeing Propulsion Systems Division	Washington	Propulsion systems engineering and procurement services
Moog, Inc.	New York	Flight control actuators
Kidde Technologies (announced May 2004)	North Carolina	Fire protection system
Toray Industries (announced May 2004)	Washington	Pre-impregnated composites
Parker Hannifin (announced September 2004)	Ohio	Hydraulic subsystem

Table 2.4—Continued

Company/Business Unit	Main Location	787 Work Statement
Monogram Systems (announced November 2004)	California	Water and waste system
Air Cruisers (announced November 2004)	New Jersey	Escape slides
Delmia Corp. (announced November 2004)	Michigan	Software
Intercim (announced November 2004)	Minnesota	Software
Korry Electronics (announced January 2005)	Washington	Flight-deck control panels
C&D Zodiac (announced April 2005)	Washington	Sidewalls, window reveals, cargo linings, door linings, and door surrounds
Securaplane (announced April 2005)	Arizona	Wireless emergency lighting system
Donaldson Company, Inc. (announced May 2005)	Minnesota	Air purification system
Astronautics Corp. of America (announced May 2005)	Wisconsin	Electronic Flight Bag (EFB)
PPG Aerospace (announced December 15, 2005)	Alabama	Electrochromic windows
Vought Aircraft Industries (announced July 2009)	Texas	Longerons, stringers, shear ties, and frame assemblies

the capability of producing metallic parts, sheet metal, machined parts, composite parts, wiring assemblies, and tubing.

By the end of the 1990s the major U.S. companies had divested their metal fabrication facilities and were procuring from other sources—primarily although not exclusively domestic. For example, Boeing St. Louis (then McDonnell Douglas) sold its metal and composite manufacturing capabilities to a British company, GKN Aerospace. In a different approach, the European consortium applied group

Table 2.5
Non-U.S. Suppliers for the 787 Dreamliner

Company/Business Unit	Main Location	787 Work Statement
Alenia Aeronautica (announced November 2003)	Italy	Horizontal stabilizer, center fuselage
Boeing Fabrication (announced November 2003)	Washington, Canada, Australia	Vertical tail assembly, movable trailing edges, wing-to-body fairing, interiors
Fuji Heavy Industries (announced November 2003)	Japan	Center wing box, integration of the center wing box with the main landing gear wheel well
Kawasaki Heavy Industries (announced November 2003)	Japan	Main landing gear wheel well, main wing fixed trailing edge, part of forward fuselage
Mitsubishi Heavy Industries (announced November 2003)	Japan	Wing box
GE Aviation (formerly Smiths Aerospace) (announced February 2004, June 2004)	United Kingdom	Common core system, landing gear actuation and control system, high lift actuation system
Eaton Aerospace (formerly FR-Hi Temp) (announced March 2004)	United Kingdom	Pumps and valves
Rolls-Royce (announced April 2004)	United Kingdom	Engines
Thales (announced July 2004, August 2004, September 2005)	France	Electrical power conversion, integrated standby flight display, in-flight entertainment system
Messier-Bugatti (announced November 2004)	France	Electric brakes

Table 2.5—Continued

Company/Business Unit	Main Location	787 Work Statement
Latecoere (announced November 2004)	France	Passenger doors
Panasonic (announced December 2004, November 2005)	Japan	Cabin services system, in-flight entertainment system
Bridgestone (announced December 2004)	Japan	Tires
Ultra Electronics Holdings (announced December 2004)	United Kingdom	Wing ice protection systems
Ipeco (announced April 2005)	United Kingdom	Flight-deck seats
Diehl Luftfahrt Electronik (announced April 2005)	Germany	Main cabin lighting
Jamco (announced April 2005, May 2005)	Japan	Lavatories, flight deck interiors, flight deck door and bulkhead assembly
CIT Systems (announced August 2005)	Sweden	Zonal drying system
PFW (announced October 2005)	Germany	Metallic tubing and ducting
Saab Aerostructures (announced October 2005)	Sweden	Large cargo doors, bulk cargo doors, and access doors
Korean Airlines–Aerospace Division (announced October 2005)	Korea	Raked wing tips for the 787-8

technology principles to organize operations across the firms.[15] Essentially, functionally grouped machines (producing parts or products with similar characteristics) were organized into cells to achieve high levels of repeatability. The consortium combined sites and created a more efficient machined-part manufacturing source, utilizing the latest technology and capital equipment while maintaining individual integrating roles.

Since then, the business model for the major aerospace companies has further evolved. Prime contractors have formed global alliances. Having divested a substantial share of their fabrication capabilities, they now use worldwide supply chains and identify their core competencies as designing, engineering, and integrating systems and platforms. Table 2.6 contrasts some of the changes between the business practices of 2000 and 2010. The primes have divested metal and composites fabrication, but they have retained software, mission systems, integration, final assembly, and checkout.

International Investment in the United States

International firms are not waiting for U.S. prime contractors to offer them subcontract work. Non-U.S. aerospace corporations are vying to enter the U.S. defense procurement arena in an aggressive manner. EADS North America has already entered the U.S. competition for aerial refueling tankers; BAE Systems and Rolls Royce have established operations in the United States. The United States Air Force (USAF)

---

[15] *Group technology* (GT) is a manufacturing philosophy in which parts having similarities (geometry, manufacturing process, and/or function) are grouped together to achieve a higher level of integration between the design and manufacturing functions of a firm. The aim is to reduce work in progress and improve delivery performance by reducing lead times. GT is based on a general principle that many problems are similar and that by grouping similar problems, a single solution can be found to a set of problems, thus saving time and effort. The group of similar parts is known as a part family and the group of machineries used to process an individual part family is known as a machine cell. It is not necessary for each part of a part family to be processed by every machine of a corresponding machine cell. This type of manufacturing in which a part family is produced by a machine cell is known as cellular manufacturing. Manufacturing efficiencies are generally increased by employing GT because the required operations may be confined to only a small cell, thus avoiding the need for transportation of in-process parts.

Table 2.6
U.S. Aerospace Business Practices, 2000 and 2010

Category	2000	2010
Metal fabrication	Domestic	Domestic and international
Composites fabrication	Primes and some domestic suppliers	Fewer primes, more domestic and international suppliers
Component assemblies	Primes and domestic partners	Domestic and international suppliers/partners
Final assembly and checkout	Primes only	Primes and selected overseas partners
Mission systems (MS)	Domestic	Domestic
Vehicle systems (VS)	Domestic	Domestic and international suppliers
Design	Primes and domestic suppliers for MS/VS	Primes; domestic and some international partners
Castings/forgings	Design by primes	Design by suppliers
Company priorities	Preserve core technologies	Return on shareholder equity and net assets
Market	Primarily domestic and some international	Increased international demand
International outsourcing	As needed offset	Increased offset and strategic sourcing overseas—best value
DoD procurement	Domestic	Low-end aircraft from overseas
Primes	Manufacturers and integrators	Integrators
Software	Primes	Primes
Design tools	CAD 3D wire	CAD solid model
Machining	Some high-speed machining	Increased high-speed machining; less tooling
Assembly	Some automation	Increased automation
Manufacturing and flight test simulation	Minimal	Increased simulation

has already started procurement of a C-27J cargo aircraft completely manufactured by Alenia Aeronautica in Italy. In a bid to win the contract to build the U.S. military's future trainer (designated the T-X), Lockheed Martin is expected to propose the T-50 Golden Eagle aircraft (jointly developed with Korean Aerospace Industries), while BAE is expected to offer its Advanced Hawk Trainer.

All the foreign aerospace companies are proposing to establish final assembly and integration of these platforms in the United States, citing creation of U.S. jobs and enhancing the U.S. aerospace infrastructure. In this concept, a significant portion of the hardware would be procured from overseas, very similar to the business model used by Japanese automakers—Toyota, Nissan, Mazda, and others.

Overseas Procurements and Technology Investments

A significant recent development has been the intent of the Government of India (GOI) to procure more than $50 billion in defense products from abroad over the next five to ten years.[16] All procurement is subject to strict offset requirements, so that 30–50 percent of the value of the buy will be produced in India.[17] Even commercial aircraft procurement by GOI-owned airlines, such as Air India, are subject to a similar 30 percent offset. India's cheap labor rate, infrastructure cost, adequate availability of technical manpower, and willingness to invest are driving all the prime aerospace companies in the West to seek Indian joint venture partners, as well as sources from which they can procure detail parts and assemblies. General Electric and Boeing have established Indian technology centers; Lockheed Martin, EADS, and Sikorsky are planning to follow.

---

[16] India is in the procurement cycle for six C-130Js, eight P-8Is, and ten C-17s, with additional follow-on options. The down-select process for 126 medium multirole combat aircraft (MMRCA) is in process; the F-16 and F/A-18E/F were contenders.

[17] Ministry of Defence, Government of India, *Defence Procurement Procedure 2008*, New Delhi, July 2008.

Potential Future Competition in the Aircraft Manufacturing Sector

A possible issue of interest to members of Congress is increased competition for the domestic industry from low-cost competitors, including the emergence of possibly strong aerospace manufacturing centers in China and Russia. Both nations appear to have plans to dominate a much larger share of their domestic markets and, in turn, perhaps the global market. China is working to develop airplanes that could become globally competitive in both the regional jet and large commercial jet aviation market, as well as in some military markets.[18] Russia has stated that it wants to become the world's third-largest aircraft manufacturer by 2015.[19] Both Chinese and Russian aircraft manufacturers face significant hurdles in building commercial aircraft, since neither has built such airplanes for the global market, which requires planes to be reliable, have low operating costs, be easily maintained, and be certifiable by authorities. A consensus view among industry watchers is that China, India, and Russia are likely to emerge as significant players over the next two decades, a development that will give Western companies major short-term cost reduction opportunities but perhaps additional long-term competitors.

Trends in Outsourcing

Generalizing from interviews with the prime contractors, the trends are unambiguous: An increasing share of work has been delegated from the prime contractors to partners and subcontractors. By some accounts, primes are outsourcing 60–80 percent of the dollar value of their contracts.

However, our interviews and discussions with prime contractors confirmed that it is not just "build-to-print" parts that are being manu-

---

[18] The Comac C919, an approximately 156-seat aircraft with dimensions similar to the A320, is in development, although a production date has not yet been announced. Additionally, in early 2011, China entered the stealth fighter realm with publication of pictures of the J-20. See Elisabeth Bumiller and Michael Wines, "Test of Stealth Fighter Clouds Gates Visit to China," *New York Times*, January 11, 2011.

[19] United Aircraft Corporation (UAC) is a joint stock company owned by the Russian government. UAC has stated it plans to become the world's third-largest aircraft manufacturer by 2015.

factured by first- and second-tier firms. These non-primes are doing more core design and assembly/integration work. Complicated mission and vehicle systems (landing gear, arresting hooks, ejection seats) have long been subcontracted, and high-performance turbine engines have been designed and produced by non-primes throughout the history of aviation.

Aerospace Workforce

The availability of a highly educated and skilled workforce is a critical component for continued innovation in the aerospace industry. The U.S. aerospace workforce is characterized by two trends: a decline in its overall population over the last two decades and a demographic profile that is heavily skewed toward an older workforce nearing retirement.

As shown in Figure 2.13, the population of the U.S. aerospace workforce has contracted considerably since 1987. Some decline can be expected given the smaller number of new start programs, lower

Figure 2.13
Aircraft Manufacturing Workforce Census (thousands), 1967–2007

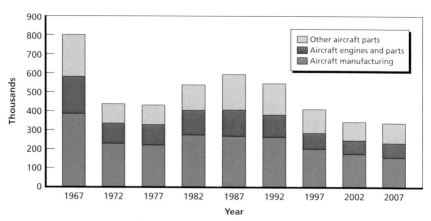

SOURCE: U.S. Census, Manufacturers and Aircraft Manufacturing.
RAND MG1133-2.13

levels of aircraft production, and improvements in workforce efficiency gained through automation and the use of computer aided design programs.

Labor Pipeline[20]

If the size of the aerospace workforce holds steady, there could be considerable demand for new talent in the coming years. The number of degrees awarded in the fields of science and engineering to students in the United States is a good indicator of the skilled labor pipeline. Science and engineering (S&E) bachelor's degrees have consistently accounted for about one-third of awarded bachelor's degrees in the United States for the past 15 years.[21] And, as shown in Figure 2.14, the total number of students enrolled in engineering programs at both the undergraduate and graduate levels has risen considerably since the mid-1990s.[22] These statistics suggest that in terms of quantity there will be an ample pipeline of educated recruits available.

Recruitment Challenges

Industry-sponsored studies have observed that competition for engineering talent from such high-tech firms as Google, Apple, and Oracle, coupled with the need to hire workers who can gain security clearances, significantly challenges recruiting efforts.[23]

---

[20] This section focuses on the portion of the workforce requiring university degrees in specialized fields. Because the touch labor workforce relies on skills for which individuals can be specifically trained in a relatively short amount of time, a potential labor shortage or attrition of skills in this area is of less concern.

[21] *National Science Board, Science and Engineering Indicators 2010,* Arlington, Va.: National Science Foundation, 2010, p. 2-4.

[22] *National Science Board,* 2010, p. 2-15.

[23] Aerospace Industries Association, *Launching the 21st Century American Aerospace Workforce,* Arlington, Va., December 2008.

Figure 2.14
Engineering Enrollment in U.S. Universities, by Level, 1979–2007

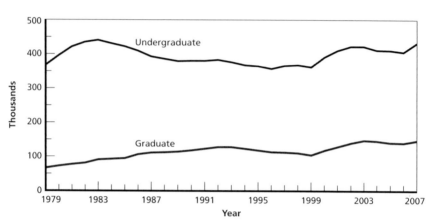

SOURCE: *Engineering and Technology Enrollments*, American Association of
Engineering Societies, Engineering Workforce Commission, various years.
NOTE: Data include full- and part-time students.
RAND MG1133-2.14

Security requirements pose a particular challenge. With a significant number of positions requiring security clearances, many non-U.S.
citizens are not eligible to be in the recruit pool or to be hired. This
especially complicates recruiting students right out of graduate school.
In 2007, the most recent year for which data are available, although
only 4 percent of S&E bachelor degrees were awarded to non-U.S. citizens, 24 percent of S&E master's degrees and a third of S&E doctoral
degrees were awarded to foreign students.[24]

Moreover, many firms that we interviewed expressed concerns
that, with new program starts spread further and further apart, new
recruits will likely work on fewer programs over the course of their
careers than their predecessors did, which ultimately could lead to a
less attractive and stimulating work environment. Countering this
trend, however, is the growing and dynamic UAS market, which may
help to attract new talent to the industry.

---

[24] *National Science Board*, 2010, p. 2-5.

Given trends in delayed retirement and a steady stream of students educated in relevant fields, there does not appear to be an imminent danger of a workforce shortage in the aerospace field. But given growing competition for the S&E labor pool, firms may have to provide greater incentives to bring new talent on board. Additionally, the future aerospace workforce will likely be less experienced, leading firms to institute better knowledge management practices. RAND did not investigate whether a less experienced workforce would have a positive or negative impact on innovation.

Implications for the Military Aircraft Industry of Changes in Organization, Funding, and Business Practices

The fixed-wing military aircraft industry's operating climate and business models continued to evolve between 2003 and 2010. As a percentage of DoD TOA, fixed-wing military aircraft programs have seen their RDT&E share increase by 45 percent and their procurement share by 10 percent since 2000. As a result, in coming years, the industrial sector involved in designing and producing these systems is likely to face increasing pressure from the U.S. government to reduce costs just as the sector is facing more competition from overseas firms. The number of primes has stabilized at three—four, if General Atomics is included. The primes' business practices since 2003 have made globalization and outsourcing more the norm rather than the exception. They now focus on providing system capabilities; as a result, they have shifted increasing portions of their design and production to sub-tier suppliers.

Although there may be a steady supply of engineers and designers to take the place of similarly trained individuals leaving the aircraft industry, career opportunities they may have in non-aircraft sectors may pose recruiting challenges for U.S. aircraft primes and for first- and second-tier suppliers.

CHAPTER THREE

Fostering Innovation in a Changing Defense Industry: What We Can Learn from Commercial Trends

At the time that RAND completed its previous study in 2003, the aerospace industry had just gone through several decades of consolidation and mergers, and policymakers expressed concerns that if further consolidations occurred the industry might jeopardize its predisposition to be innovative. This chapter discusses trends in innovation in the aerospace industry and in industry in general that have taken place since the 2003 study.

As we discussed in Chapter Two, the aerospace industry has changed over the past decade such that prime contractors are passing more and more work down to second- and third-tier partner firms. This, in turn, has changed where and when in the industrial cycle innovation occurs. It also has implications for how the government, working with prime contractors, can create and nurture environments that most effectively foster innovation.

The relationship between innovation and this emerging aerospace industry model has not been extensively studied, and researchers have little quantitative data to gauge whether today's environment is more conducive to innovation, or less so, than it was in past decades.

Because there have been few studies focusing on innovation in military systems, in this chapter we look at analogous industries outside of defense aerospace that have similarly turned to sub-tier partners to strengthen their competitive and innovative positions.[1] These non-

---

[1] Many examples of innovative practices concern companies that make relatively simple products: drugs, iPods, consumer staples, and the like. These products are nothing like military fixed-wing aircraft in terms of complexity. Moreover, the commercial world is charac-

defense industries may provide insights on ways to foster competition and innovation in the U.S. defense aerospace industrial base.

There is little doubt that there have been high levels of innovation in the defense aerospace industry, as exemplified by stealth, avionics, and other technologies. Rather, the issue is whether the innovation pattern in aerospace is similar or dissimilar to other industries. If it is similar, then policymakers may want to consider "borrowing" or "transferring" concepts, management ideas, or distinctions from other industrial sectors whose business models have evolved in similar fashion.

The insights in this chapter, therefore, are not taken from any one industry. They are taken from multiple sectors and transcend any single industry. They also avoid focusing on the aerospace industry's preoccupation with such immediate pressures as keeping costs down and performance up. These insights include

- ways to increase innovation
- new insights on innovation dynamics
- problems that have appeared in other industries that are now appearing in the aircraft industry—especially coordinating complex networks of suppliers
- lack of integration between R&D and the strategies of firms. R&D can, for example, be driven by R&D budgets, not the demands of customers
- the finding that long-term success in industries characterized by technological developments often comes from superior knowledge rather than bargaining power and corporate positioning— the factors that are traditionally emphasized in study commis-

---

terized by customer bases that number in the tens of thousands, if not millions. In contrast, military aircraft producers have few customers, often one. But those few customers constitute a hierarchy of decisionmakers with different perspectives about the ultimate product and how it is acquired. That hierarchy includes Congress, DoD, the military services, the operating commands, and the users in the field. This array of disparate interests doubtless has an influence on the degree of innovation, one that differs significantly from the commercial world, which is not as layered and has multiple interests with ownership stakes in programs. In addition, the commercial world's focus on private-sector customers as opposed to policy- and national security–oriented military customers likely has some influence on innovation.

sions. For example, new research findings suggest that increasing the number of suppliers alone may have little impact on innovation or even on cost control.

The discussion in this chapter is organized around three concepts:

- The changing locus of innovation in many industries, that is, where innovations come from. This is a relative shift—away from large mature companies to small and medium-sized companies. It creates a more complicated value chain: The activities that go into new products (innovation, design, production, distribution, servicing, and so forth) cut across more departmental boundaries and now reach outside of the firm.
- The growing recognition that managing production and innovation networks is central in technology-based businesses, and that this type of management is different from the traditional hierarchical approach of the past.
- The recognition of the critical role of risk capital in innovation, specifically, how external sources of capital (venture capital, private equity) complement traditional sources of capital from DoD.

The Changing Locus of Innovation

The locus of innovation in the U.S. economy has fundamentally changed.[2] The trend is for innovation to occur in smaller companies, often geographically concentrated into clusters, with access to venture capital backing.[3]

As projects have become larger in scale and more complex in technology, the locus of knowledge and knowhow has shifted to small- and medium-sized firms. Large firms still exist and remain important. But they are not able to harness all the skill sets that they did in past

---

[2] For an overview of changes in innovation see William J. Holstein, *The Next American Economy*, New York: Walker & Co., 2011.

[3] Holstein, 2011.

decades. Therefore, their role has shifted to integrating these smaller and medium- sized suppliers in efficient ways. This change represents a very different kind of management that is only now being recognized in business schools and in the literature.

An illustrative example shows this trend. IBM once dominated the computer industry. It could internalize whatever skills it wanted by hiring the right people. But information technology (IT) today is far too complex for any one company to do this. Microsoft, Google, Cisco, Hewlett Packard, and now social networking companies offer a much wider range of system design options than existed in the past. These large companies reach out to smaller firms for many of their innovations. This reaching out can take many different forms. It includes scanning the horizon outside the company for innovative organizations, outsourcing innovation to them, working in collaborative partnerships, and acquisition. The exact form of this reaching out varies by industry, large company strategy, and the details of the technology in question. However—and this is the main point—reaching outside the firm is increasing in most technology-intensive industries. This is true to such an extent that there is now a literature on ways to reach out and ways for a large company to protect itself from not being intellectually hollowed out.[4]

IBM has to pick and choose what parts of the value chain it is going to compete in-house. Any goods and services that it does not produce in-house, it must buy on the market. This means cooperating with other firms and competing with them at the same time. IBM's relations with smaller firms, the ones who have critical skills that IBM needs for its overall corporate strategy, thus become absolutely central to IBM's success.

This shift to managing innovations in smaller and medium-sized enterprises has large implications for the national system of innovation in the United States.[5] Innovation used to reside in the large corpora-

---

[4] For one example see Richard Leifer et al., *Radical Innovation: How Mature Companies Can Outsmart Upstarts*, Boston, Mass.: Harvard Business School Press, 2000.

[5] The essential framework for analyzing this discussion can be found in Richard Nelson, ed., *National Innovation Systems*, Oxford, UK: Oxford University Press, 1993. It should be

tion. They were the only ones who could afford the R&D, industrial laboratories, and the costs of intellectual property protection in the form of patents and legal instruments.

It is not accurate to say, however, that the small enterprise has eclipsed the larger corporation. Indeed, competition is also driving significant innovation in large American companies. A better way to state what has occurred is that innovation is now spread over many firms. It can turn up in a company of any size. The wider availability of capital, the rise of "Silicon Valleys" in several geographic regions, and other factors converge to broaden the locus of innovation. The robotics industry has taken hold in Pittsburgh, computer simulation in Orlando, and genomics in San Diego. This trend is now going beyond the scale of the national innovation system. True global innovation systems are emerging. Companies have technical centers, and R&D, all over the world.[6]

What are the implications of the trend toward knowledge, skills, and performance expertise increasing in smaller and mid-size companies? There are several. First, it is dangerous for a large company to rely exclusively on in-house R&D as the only source of innovation. Doing so is likely to miss important opportunities because of inadequate knowledge inside the firm. It is also too slow. Richard Foster, former senior director at McKinsey & Company, makes the point in the following way:[7] The market, he says, is more innovative than any company in it. This sounds at first like a trivial insight. But as the number of small firms increases, so does the number of new value ideas. The large firm cannot possibly keep up with technology through in-house R&D alone.

Large companies have developed management frameworks to search for innovations wherever they may be found. IBM and GE, as an example, maintain R&D centers overseas to keep tabs on what is

---

noted, though, that the pace of change has accelerated far beyond this framework of the early 1990s.

[6]   See Bruce McKern, ed., *Managing the Global Network Corporation*, London: Routledge, 2003.

[7]   Richard N. Foster, *Innovation: The Attacker's Advantage*, New York: Summit Books, 1986.

going on and to break away from the narrow perspective of their in-house R&D staff.

Second, companies have developed rapid learning strategies to quickly come up to speed in an area of interest. These strategies range from accelerated executive education programs to investing in private equity (PE) funds as a way to get a window on a technology sector. The average length of executive education programs has shrunk, in part because companies find it too slow to go through lengthy bottom-up introductions to new technical areas. Many companies invest in PE funds to get seats on the board of directors of technology companies that the PE fund invests in. Such strategic investing is done not for the narrow purposes of generating a large return on investment, but rather to learn about a new technology area.

A third implication is a growing acceptance that there are more-complex forms of innovation. Small firms may work with large ones in a flexible way. Clusters of innovation (see above) work as a free market without central direction. New types of contracts are written to protect the rights of the less powerful actors in this new landscape. The flexible nature of network relationships—from working arrangements to intellectual property protection—is a distinctive feature of the new innovation economy.

Fourth, the organizational structure of the large firm itself is changing. The large firm is likely to be disaggregated into smaller teams. The large "org chart" of the past may bear little relationship to communication and information flows as they actually work inside a large company today. Teams are where the real innovative work of the firm is carried out.

One interesting question is whether these teams differ from the integrated product teams (IPTs) used in the defense industry for the past decade. Our sense is that there are major differences. Teams in networked industries are distinguished by whether they are "I teams" (with an internal focus, e.g., on corporate processes) or "X teams" (with an external focus, e.g., on new technologies and markets). X teams are used like scouts, plotting the companies' path in the innovation land-scape. Apple, for example, uses X teams to explore how new technologies are used by actual customers.[8] Customers make up part of Apple's X teams.

Our sense is that these kinds of teams are very different from DoD's IPTs. Although the primary motivation behind IPTs is to get several disciplines working together at the task level, IPTs often focus on compliance, on adherence to regulations. And these regulations originate not from some flexible market, but from authoritative sources (Congress and DoD). This appears to be the very opposite of a decentralized approach to innovation.

Like any landscape, the new innovation landscape needs to be mapped. In this approach, mapping is used more for exploration than for optimization. It is, once again, the near opposite of a Planning, Programming, and Budgeting System (PPBS) approach to innovation. Mapping exercises are how companies "see" and "classify" the terrain ahead. Some innovation regions may be viewed as high risk. Other regions are seen as overpopulated—congested with too much attention and risk capital. Still others are unexplored. And some regions of the innovation landscape complement each other.

This shift in the locus of innovation in the U.S. economy leads to two broad observations: First, innovation is more likely to be outside of the managerial and legal control of a firm. This is why innovation networks have become so important. Second, spotting innovations has become more difficult. Thus, new sources of risk capital have been created not merely to produce higher returns for investors but also to reconnoiter the innovation landscape, something that a company often does not have time to do itself. We turn now to these two topics.

---

[8]   Stefan Thomke and Barbara Feinberg, *Design Thinking and Innovation at Apple*, Harvard Business School Case 9-609-066, March 4, 2010.

Managing Innovation in Networks

The past decade has seen a great deal of attention and research in business schools on managing innovation in networks.[9] There are two principal reasons for interest in the topic:

- Companies, from small to large, are becoming more network-like in their organization.
- Complex projects require a large number of cooperating enterprises working together, and this demands a very different management style.

The "old" corporate model was a pyramid organized as a layered hierarchy. Those in the top layers told those below them what to do. Innovation tended to be mostly in house, e.g., in industrial laboratories or in R&D subdivisions. Strategy for the whole company was crafted at the top and communicated to those below. There was little lateral communication or coordination horizontally by divisions because each silo tended to its own market. In terms of innovation and risk-taking, there tended to be little that did not come out of "official" innovation channels. Those lower in the hierarchy had little authority—or budget—to take risks or innovate. Their job was to follow processes, rules, and regulations about the flow of work and information.

This corporate model is increasingly disappearing. Market turbulence, new global competitors, and new technologies, most especially IT, have made it less agile and have underscored the disadvantages of bureaucracy. To be clear, however, this old model is not a caricature of the past. Rather, it is a successful model for a corporation in a stable environment. If market conditions do not change too much in any one year, it can be, and has been, highly successful. General Motors and (the old) IBM are examples of successful process-driven bureaucracies.

---

[9] See Ranjay Gulati, *Managing Network Resources, Alliances, Affiliations, and Other Relational Assets,* Oxford, UK: Oxford University Press, 2006; and John Roberts, *Organizational Design for Performance and Growth,* Oxford, UK: Oxford University Press, 2006.

But as markets became more turbulent as a result of technological change and other factors, performance has declined.[10] The new network model is based on smaller teams with much greater autonomy—and larger budgets—a delayering of authority structures and processes and a great emphasis on agility. In short, the new structure looks more like a network.

Much of what a company once produced is now outsourced. It is purchased on the market, illustrating the first trend we discussed, the changing locus of innovation. A critical strategic decision today is whether to purchase something on the market or to develop it in house. Getting this right is increasingly judged to be a major strategic advantage.

The second trend is toward reconceiving management as managing "networks" distinct from managing plants, offices, or people. Pharmaceutical companies, for example, have shifted anywhere from 50 to 75 percent of their R&D outside the firm.[11] They invest in smaller start-up companies that are closer to the cutting edge of innovation and that are less encumbered by the results and profit concerns that color larger organizations.

Apple is a good example of the trend. It buys digital signal processor chips, light emitting diodes, and microphones from China or other low-cost suppliers. Apple itself makes little of its new i-product lines. Its strategy is to use its own operating systems as the central architecture around which these components are organized.

Procter & Gamble (P&G) provides another example of innovation conceived as a network task. P&G uses a rule of thumb that 70 percent of its innovations should originate outside the company.

---

[10] This is a fundamental finding in complex organization theory. See Charles Perrow, *Complex Organizations,* New York: Random House, 1986, especially Chapter 4.

[11] See Oliver Gassman, Gerrit Reepmeyer, and Maxmillian von Zedtwitz, *Leading Pharmaceutical Innovation: Trends and Drivers for Growth in the Pharmaceutical Industry,* Berlin: Springer, 2008.

The idea is that P&G needs to be in a larger innovation network than could possibly be created inside the company.[12]

The in-house industrial lab, the Chief Executive Officer's (CEO's) brain trust, etc., are declining sources of new ideas for products and services. Studies increasingly show that innovation occurs in the perimeter space of a firm—even outside the zone of companies that it monitors on a regular basis.[13] Innovations may come from firms that are several steps removed from the firm looking for novelty and value.

This means, in turn, that management processes need to be created or modified to make sure all of this happens in a timely way. A key message here is that this is a managed process, not an ad hoc one. Otherwise, valuable innovations will never make it to the inner core of the company where they will finally be reviewed for decision.

Reconceiving innovation as a network task has several managerial consequences:

- **"Steering" Innovations to the Core.** Innovations coming from unusual or nonstandard sources risk being killed off because they are not made in-house. They need to be tracked, evaluated, and steered to the core of the company, where the ultimate go–no go decision is made.
- **Cross-Boundary Threshold Problems.** The most dangerous boundary in many companies is the one between departments. It is a danger posed by almost all new disruptive technologies, according to Professor Clay Christensen of Harvard Business School. While some scholars disagree with this pessimistic conclusion, there is little doubt that it is, nonetheless, a fundamental divide. The advice here is to use a different set of metrics for the innovations than are used for ordinary products. For example, they may be allowed to have lower performance standards (as long

---

[12] This is the central argument in Mark W. Johnson, *Seizing the White Space: Business Model Innovation for Growth and Renewal*, Boston, Mass.: Harvard Business Press, 2010, which contains an introduction by Procter & Gamble CEO A. G. Lafley.

[13] See Joel Podolny, "Social Capital," briefing slides, New Haven, Conn.: Yale School of Management, 2007.

as they are rapidly improving), and they may be protected by different financial metrics, e.g., lower return on investment hurdles.
- **Speed and Agility.** Finally, speed and agility in managing this systematic innovation process is critical. The world in which a scientist puttered in the lab for years and came up with something new is rapidly disappearing. If support for basic research is declining, a scientist probably will not keep his or her job long enough to make a difference.

It is useful to step back and ask whether the military aircraft industry has followed the patterns described here for managing innovation in networks. The aircraft industry was a pioneer in the 1950s and 1960s in the network model of multiple-tier suppliers. It was, indeed, a networked industry. This is often pointed out in descriptions of the industry.[14]

However, there is also reason to believe that innovation in military aircraft is not keeping pace with modern management developments. A widely held view among individuals we interviewed is that military aircraft companies have a culture in which innovation at lower levels is discouraged. "Leading from the middle" management, characteristic of the industry in its formative years, does not appear to characterize it today.

Regardless of the perspective an observer takes on the aircraft industry, key insights emerge from research on innovation in networks, insights that are important to this sector. This research has found that a company does not—indeed, cannot—transform itself overnight into a networked organization. But smaller steps can be taken to move in that direction.[15] These include

---

[14] For histories of the U.S. defense industry that describe its structure, see Jacques S. Gansler, *The Defense Industry*, Cambridge, Mass.: MIT Press, 1982; and the classic volume by Merton J. Peck and Frederic M. Scherer, *The Weapons Acquisition Process: An Economic Analysis*, Boston, Mass.: Harvard Business School Division of Research, 1962.

[15] A good review of this development is in Ranjay Gulati, "Silo Busting, How to Execute on the Promise of Customer Focus," *Harvard Business Review*, May 2007; also see Ranjay Gulati, *Reorganize for Resilience, Putting Customers at the Center of Your Business*, Boston, Mass.: Harvard Business Press, 2009.

- processes that transcend, rather than destroy, the existing hierarchical structures of the firm
- creation of boundary-spanning roles
- educational programs to increase knowledge in divisions and disciplines about their respective activities
- developing an "outside in" corporate culture.

These are small, incremental changes. Cisco, for example, uses a system of internal customer champions to advocate on behalf of external key customers. A "champion" has the authority and knowledge to cut across departments to demand solutions for an external customer. But Cisco did not obliterate its existing marketing and services departments. Rather it appended the customer champion groups to them.

Platform-Mediated Networks

A more advanced kind of network innovation is through platform-mediated networks. This is the cutting edge of innovation in the United States today. Platform-mediated networks are found in such diverse industries as credit cards, health maintenance organizations (HMOs), travel reservations, video games, container shipping, music, and (remarkably) cement. The common feature in these sectors is that a "platform" is used to organize the way that buyers and sellers interact. An example is Apple's iTunes. It is a software platform that is replacing the (nonplatform) strategy of selling CDs in retail stores. In other words, a platform-mediated network is replacing a nonplatform-mediated strategy.

Not all platforms involved software. Or, at least, they involve much that is not software. Cemex, Mexico's global cement giant, has built its advantage around a platform that involves IT and uses GPS locators on centrally dispatched cement trucks. In the past, constructions sites were charged heavy penalties for last-minute rush orders because cement is a highly perishable commodity. But Cemex's new platform makes it possible to smooth out demand by integrating the geographic database of truck locations (via GPS) with those of construction sites. It is a different business model built around the platform-mediated interaction of customer and supplier.

Banking is increasingly becoming a platform-mediated business. Customers have accounts that cross many old-style banking boundaries. Brokerage, mortgage, checking, savings, and credit cards are integrated around a major platform. This banking platform is even penetrating into mobile telephones. Customers demand access to their various accounts wherever they are. It would be virtually impossible for an old-style bank without such a platform to compete.

This platform-centric approach is having profound implications on innovation in banking. Indeed it is hard to imagine this sector without explicitly considering the platform. It is also hard to think of bargaining power outside without the platform: Whether the bank or the customer has the bargaining power is shaped by the platform. For example, a customer may be so tied into a bank's specific platform that the deterrents to exit are large.

Today, it is estimated that 60 of 100 of the world's largest companies earn most of their revenue from platform-mediated networks.[16] Business school research points to the central observation that there is a cross-cutting similarity (namely, "platform-centric-ness") to what American Express, Cemex, Cisco, Citgroup, Intel, NTT (Japan), UPS, and Vivendi (France) are doing.

Risk Capital and Innovation

The role of risk capital in innovation is receiving a great deal of attention in economics and business research. This is because of the mushrooming growth of venture capital and private equity in the American economy, and also because of the successes they have had in backing companies that have grown to world-class status. There likely has been a slowdown in the private equity industry resulting from the 2008

---

[16] Thomas R. Eisenmann, *Managing Networked Businesses: Course Overview for Educators*, Harvard Business School Case 5-807-104, Revised, January 2, 2007.

recession. However, there is little doubt that private equity remains a formidable force, as any reading of the business press will show.

Risk capital in defense, however, receives little attention or acknowledgement outside of a small set of companies and investment banks. The general DoD perspective is still the classic linear model of defense innovation. In this model, DoD provides a certain percentage of the dollar value of acquisition contracts to defense companies to underwrite R&D. When this type of innovation began in the 1950s and 1960s, it was called independent R&D (IR&D). It had few restrictions, and was highly valued by companies in the defense sector.

Over time however, many restrictions were applied, significantly limiting its value. In addition to IR&D, the linear model entailed special programs in which military or DoD agencies (Office of Naval Research, Defense Advanced Research Projects Agency, Director, Defense Research & Engineering) awarded competitively bid directed research contracts in focused areas of interest.

The classic model in defense is changing however. Today, the Central Intelligence Agency (through In-Q-Tel) and virtually all the military services have private equity and venture capital arms.[17] The amounts in these programs are quite small, however. We could find no data on the total funding for the military services, DoD, and the intelligence community. This shows the gaping data gaps in this area.

A more important change concerns private-sector investment in innovative defense companies. This includes private equity funds, both those that specialize in defense and those that do not. It also includes strategic investment by large defense companies in smaller ones in order to obtain intellectual property, skills, and other innovations.

All of these examples can be seen as a nonlinear model of defense innovation, in that they do not involve a straight-line allocation of funding from DoD to the innovating firm.

---

[17] For the purpose of clarity of terms, *private equity* (PE) is a pool of money invested in companies to bring their business to a new level, e.g., to develop a new technology, expand to a new market. *Venture capital* (VC) is the same thing, except that it implies an earlier stage of development. VC-backed companies need not have any revenue stream from sales, whereas PE-backed companies generally do. The distinction between VC and PE, however, is not hard and fast.

The appearance of nonlinear elements of defense funding and innovation raises three questions that have received little attention as yet:

What is the true budget for defense innovation? To our knowledge, there has been no audit of the size of VC/PE investments in defense innovation. The appropriate categories have not been defined, the data have not been collected, and comparisons have not been made with "classic" DoD R&D funding. There should be scorecards for such metrics, to get an overall sense of these investments.

Does DoD need better understanding of U.S. capital markets? Companies in the defense sector, from large to small, depend on capital markets. Yet the perception is that DoD is the only capital market that matters. The treatment of Lockheed or Northrop in the capital markets, including their share price, has an increasingly decisive effect on their corporate strategy. Study after study has concluded that the big defense companies are influenced by Wall Street. Yet the links between them and the capital markets are almost never drawn. It would be useful for DoD to have a better understanding of the relationship between capital markets and innovation.[18]

Is DoD in a unique position to improve innovation? The strong, quite understandable tendency is for DoD to interact with the larger defense companies. Smaller companies tend to interact with DoD through regulations and small-business set-asides. Medium-sized companies vary between these two poles. It seems to us that DoD has no overall design for increasing innovation that acknowledges the changes taking place in the U.S. economy. Moreover, DoD appears to have no framework for what such a policy should look like.

Yet DoD as a monopsony (that is, a single buyer) is in a unique position to shape innovation in the defense industry. Often, DoD's single-buyer aspects are considered a disadvantage. The argument is made that the specialized nature of defense products justifies the cost because suppliers have no other markets to turn to. However, if this argument

---

[18] The literature on capital markets and innovation is considerable. See Mariana Spatareanu, "The Cost of Capital, Finance and High-Tech Investment," *International Review of Applied Economics*, Vol. 22, No. 6, November 2008, pp. 693–705.

were valid it would make defense a truly unique industry, one whose dynamics are not affected by the factors that determine outcomes in computers, telecommunications, pharmaceuticals, transportation, retail, and so forth. A monopsony has unique power to shape industry structure. That DoD does not do so in the U.S. defense industry is a matter of choice, not one having to do with the specialized nature of the business. The cost of this choice is rising steeply, as measured by the unit cost of new airplanes, information systems, and munitions.

An organization like DoD that does not take account of future innovation is hurting its own future. There are major public policy reasons for taking a shaping role in the defense industry, but these do not appear to have been important factors shaping DoD industrial policy on the sector. Whether defense innovation can move into the future in the same way that other sectors (IT, banking, etc.) have done is hardly clear. The subject merits much more sober thinking than it has received.

CHAPTER FOUR

Prospects for Innovation and Competition in the Fixed-Wing Military Aircraft Industry: Programs of Record and Alternative Future Programs

As we have seen in Chapter Two, the 2010 snapshot for the fixed-wing military aircraft industrial base shows that the shares of work going to the primes have become unbalanced. Although procurement funding is spread out fairly evenly among primes, RDT&E funding in 2010 mostly went to a single company. This chapter will examine options to create a more balanced funding distribution so as to satisfy Congress' criterion of having at least two healthy primes with equal shares of RDT&E and procurement.

Program of Record/Base Case FY 2010–2025

We used DoD's FY 2011 program of record—the roster of programs that have survived the Program Objectives Memorandum (POM) process[1] and are listed in the Future Years Defense Program (FYDP)—to

---

[1] According to the Defense Acquisition University, the POM "is an annual memorandum in prescribed format submitted to the Secretary of Defense (SECDEF) by the DoD Component heads, which recommends the total resource requirements and programs within the parameters of SECDEF's fiscal guidance. The POM is a major document in the Planning, Programming, Budgeting and Execution (PPBE) process, and the basis for the component budget estimates. The POM is the principal programming document that details how a component proposes to respond to assignments in the Strategic Planning Guidance (SPG) and Joint Programming Guidance (JPG) and satisfy its assigned functions over the Future Years Defense Program (FYDP). The POM shows programmed needs six years hence." See Defense Acquisition University, ACQuipedia, Program Objective Memorandum (POM/ Budget Submit [on-year]), May 2010.

derive our baseline for RDT&E and procurement funding. Figures 4.1 and 4.2 portray the baselines for those funding elements. Figure 4.1 shows the RDT&E baseline, which identifies future RDT&E funding. Note that there is no funding in the baseline identified beyond FY 2018. Figure 4.2 depicts the procurement baseline, which identifies future procurement funding to FY 2025. Readers should note that funding for the new KC-X tanker's RDT&E is already in the FY 2011 FYDP, whereas procurement funding for that aircraft has yet to be decided.[2]

Figure 4.1 shows that under the current funding picture RDT&E funding peaks in FY 2010, after which it falls off dramatically. This suggests that a lack of new starts will create a bleak future for design, planning, and development capability in the second half of the decade. In the opinion of the individuals we interviewed, there may not be enough RTD&E funding to sustain a vigorous technology and design/development staff at all primes. The figure also shows that the F-35 and

Figure 4.1
Program of Record, RDT&E Funding, FY 2000–2025

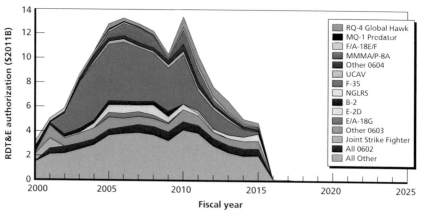

NOTES: 0602, 0603, and 0604 are R&D categories that refer to technology readiness. NGLRS = Next-generation long range strike bomber. MMMA = Multimission Maritime Aircraft.

RAND *MG1133-4.1*

---

[2] We recognize that RDT&E will not cease in FY2018, but the figure reflects the most recent data available.

All Other categories dominate RDT&E funding today and that All Other will dominate in FY 2015.

Figure 4.2 shows that under the current funding plan many procurement programs will end around FY 2015 or shortly thereafter, and that funding will effectively drop by half for several years before dropping further. Moreover, without the F-35 planned procurement, the picture is even bleaker. This suggests that the current program cannot sustain basic infrastructure and business base for primes.

The RDT&E and procurement program of record funding shown in Figure 4.1 and Figure 4.2 translates into an inventory of new air vehicles that DoD plans to acquire in coming years. In Appendix D we display total quantities of air vehicles that will be acquired in the FY 2012–2021 period as outlined in the *Aircraft Procurement Plan Fiscal Years (FY) 2012–2041* that DoD submitted to Congress in April 2011.

Figure 4.2
Program of Record, Procurement Funding, FY 2000–2025

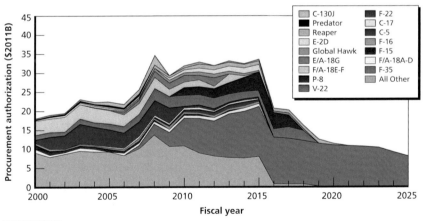

RAND *MG1133-4.2*

Altering the Program of Record/Base Case, FY 2010–2025

As we discussed in Chapters One and Two, an innovative, competitive industrial environment as spelled out by Congress is one that consists of at least two full-service prime contractors having approximately equal shares of *both* RDT&E funding and procurement funding.[3] As shown above in Figures 4.1 and 4.2, the industry environment under the current program of record will not satisfy that criterion after about FY 2015.

Some Alternative Future Programs

The contents of the current budget and SARs represent only today's outlook for the future. Experience shows that our ability to make such predictions is imperfect, and that the content of those future years will turn out to be different from today's projections. In particular, it is almost certain that some new programs will be started within that time period, and that the funding streams now forecast for current programs will change. Therefore, we need to explore some alternative future scenarios and how they might affect the military aircraft industry.

After discussions with contractor personnel, DoD officials, and colleagues at RAND and review of the Aircraft Investment Plan (FY 2011–2040), we postulated four scenarios of future aircraft development and procurement programs, starting with programs that we believe are highly likely to start in the next few years, and then extending to more speculative scenarios. Each scenario represents different dollar values and different kinds of design and development work, allowing us to get some idea of how those two parameters might inter-

---

[3]   As we noted in Chapter One, although we recognize that primes also engage in tasks, activities, and other elements that they need to share, we used funding as a quantifiable measure on which to base our criterion. While the congressional language motivated us to split RDT&E and procurement funding equally between two primes, it is not clear exactly what the shares should be. Depending on the circumstances, unequal divisions of funding—say 60:40, 70:30, or even 80:20—may be sufficient to sustain multiple primes for a period of time. Additionally, there may also be circumstances where funding could be split among three primes, either equally or unequally. However, if sustained over the long term, such unequal divisions may put lesser-funded primes at a disadvantage.

act to support the industry in different time periods. The general scenarios we examined were as follows:

1. Acquisition (RDT&E plus procurement) of three new programs: the T-X trainer, the KC-X tanker, and the unmanned carrier-launched surveillance and strike aircraft (UCLASS).
2. Acquisition of four new programs: Scenario 1 plus the F-22 fighter sold as foreign military sales (which we term F-22 FMS).
3. Acquisition of five new programs: Scenario 2 plus the next-generation bomber.
4. Acquisition of six new programs: Scenario 3 plus the sixth-generation fighter.

For each program, we

- specified start dates, program durations, and production quantities
- estimated RDT&E and procurement funding
- overlaid resource profiles on top of the Program of Record, which we depict in Figures 4.1 and 4.2 and which we refer to as the Base Case
- assessed the likely impacts on industry.

Table 4.1 details the attributes and schedules of the six programs. For all programs except the F-22, the values in the table represent what the programs may entail but are not in any sense official values.[4] They are intended to illustrate the potential impact on the three manned fixed-wing military aircraft prime contractors. The F-22 values are based on a recent RAND investigation of issues connected with terminating the F-22 program.[5]

---

[4] Note that the KC-46A tanker (KC-X) contract was awarded after we completed this analysis but before this document went to press.

[5] Obaid Younossi, Kevin Brancato, John C. Graser, Thomas Light, Rena Rudavsky, and Jerry M. Sollinger, *Ending F-22A Production: Costs and Industrial Base Implications of Alternative Options*, Santa Monica, Calif.: RAND MG-797-AF, 2010.

Table 4.1
Potential New Aircraft Program Attributes and Schedules

System	T-X	KC-X	UCLASS	F-22 Foreign Military Sales	Next-Generation Bomber[a]	Sixth-Generation Fighter
EMD ($ billion)	1.5	3.6	4	1.62	36.1	35.7
EMD start	2012	2011	2013	2012	2012	2016
EMD end	2015	2017	2018	2016	2020	2024
Procurement quantity	450	179	142	40	100	200
Average procurement unit cost ($million)	30	196	75	273	750	203
Procurement start	2016	2016	2018	2016	2020	2024
Procurement end	2025	2025	2025	2019	2025	2035

[a]The Next Generation Bomber program was formally cancelled in 2009 by Secretary Gates. He subsequently directed the development of the new penetrating Long Range Strike-Bomber. The numbers used here are notional and derived from B-2 experience. Our objective here is to give the reader an idea of how a large program affects the industrial base.

Implications for RDT&E and Procurement of Adding Three Programs (T-X, KC-X, and UCLASS) to Baseline

Using the 2011 FYDP as our funding base case, we found that if DoD were to award three new programs—T-X, KC-X, and UCLASS—to a single firm, the industry still would become uncompetitive after 2015. Figures 4.3–4.6 show that adding the three programs would not change the RDT&E base case picture significantly, nor would it appreciably change the procurement picture. As shown in Figures 4.4 and 4.6, the three programs were all assigned to Boeing, the contractor

**Figure 4.3
RDT&E Funding: Base Case Plus T-X and UCLASS Programs,
FY 2000–2025**

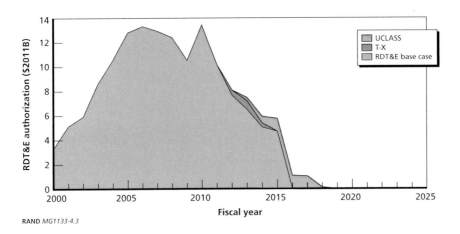

**Figure 4.4
Potential Prime Contractor Shares of RDT&E Funding: Base Case Plus T-X
and UCLASS Programs, FY 2010–2025**

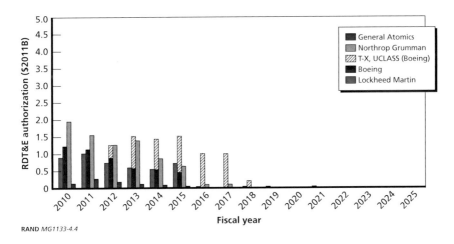

Figure 4.5
Procurement Funding: Base Case Plus T-X, KC-X, and UCLASS Programs,
FY 2000–2025

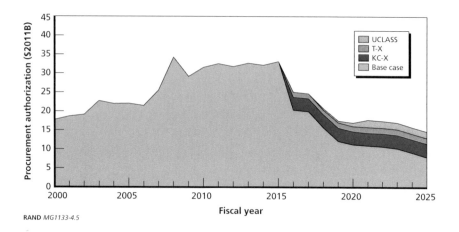

Figure 4.6
Potential Prime Contractor Shares of Procurement Funding: Base Case
Plus T-X, KC-X, and UCLASS Programs, FY 2010–2025

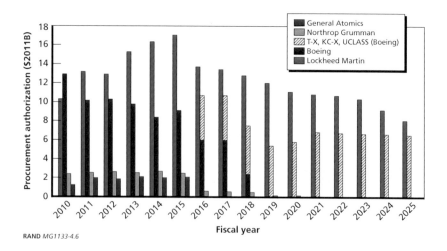

more in need of immediate funding than Northrop Grumman or Lockheed Martin.[6]

Readers who compare the current report and the 2003 document should be aware of a fundamental difference between the two studies. The previous study used SARs, contract data, and available contractor reporting data to estimate the allocation of total obligation authority between prime contractors when two (or more) were involved in specific programs (e.g., F/A-18, F-22, JSF/F-35, etc.). Because these numbers change over time, such estimates were not made for the current study. Consequently, charts depicting contractor funding levels and shares are not comparable between the two studies.

Implications for RDT&E and Procurement of Adding Four Programs (T-X, KC-X, UCLASS, and F-22 Foreign Military Sales) to Baseline

The above outcomes would also apply if, in addition to those three programs, DoD were to pursue foreign military sales of the F-22. Adding foreign military sales of the F-22 would do little to improve the RDT&E picture. As can be seen in Figures 4.7 and 4.8, foreign military sales of that aircraft would have an effect for only four years, and the effect would be felt in large measure by Lockheed Martin.

Moreover, as Figures 4.9 and 4.10 suggest, the procurement picture also would show little improvement from the addition of foreign military sales of the F-22. As with RDT&E, Lockheed Martin would be the predominant beneficiary of additional procurement funds related to the sales, and those benefits would be concentrated in the 2016–2019 period, as shown in Figure 4.10. (See Appendix A for further discussion of the implications of the sale of F-22 fighters to non-U.S. customers.)

---

[6] Note that the KC-X does not appear in Figure 4.5 and Figure 4.6, which display RDT&E data, but appears in Figure 4.7 and Figure 4.8, which display procurement data. The reason for this is because funding for the new tanker's RDT&E is already in the FY 2011 FYDP but procurement funding has yet to be decided. This differential pattern is repeated throughout this chapter's RDT&E and procurement data displays.

Figure 4.7
RDT&E Funding: Base Case Plus T-X, UCLASS, and F-22 Foreign Military Sales Programs, FY 2000–2025

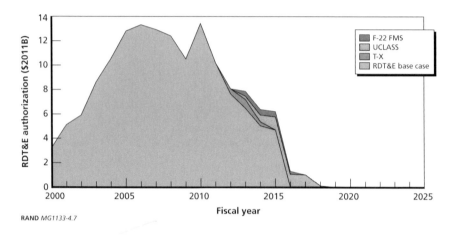

RAND MG1133-4.7

Figure 4.8
Potential Prime Contractor Shares of RDT&E Funding: Base Case Plus T-X, UCLASS, and F-22 Foreign Military Sales Programs, FY 2010–2025

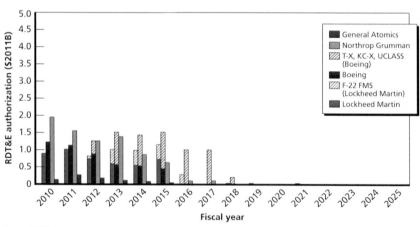

RAND MG1133-4.8

Figure 4.9
Procurement Funding: Base Case Plus T-X, KC-X, UCLASS, and F-22 Foreign Military Sales Programs, FY 2000–2025

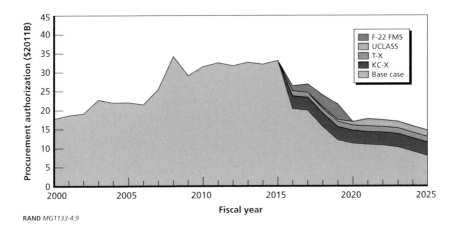

RAND *MG1133-4.9*

Figure 4.10
Potential Prime Contractor Shares of Procurement Funding: Base Case Plus T-X, KC-X, UCLASS, and F-22 Foreign Military Sales Programs, FY 2010–2025

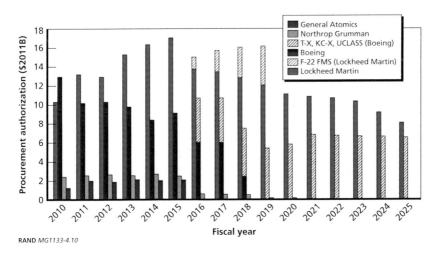

RAND *MG1133-4.10*

Implications for RDT&E and Procurement of Adding Five Programs (T-X, KC-X, UCLASS, F-22 Foreign Military Sales, and Next-Generation Bomber) to Baseline

However, involving two primes equally in performing RDT&E and procurement on a next-generation bomber would sustain two firms with approximately equal RDT&E and procurement funding. As can be seen in Figures 4.11 and 4.12, the next-generation bomber would have a significantly larger impact on the RDT&E base case than the other programs and would sustain two primes through FY 2020. As Figure 4.12 shows, if Boeing and Northrop Grumman were to share the next-generation bomber, they would receive the bulk of RDT&E funding over the next decade.

Additionally, such a strategy could reverse a decline in procurement funding and sustain two firms if each were awarded equal shares of that work. We display these results in Figures 4.13 and 4.14, againusing the 2011 FYDP as our funding base case. As Figure 4.14 shows, in this scenario—in which Boeing and Northrop Grumman would equally share next-generation bomber procurements—Northrop Grumman's procurement outlook would improve the most.

Figure 4.11
RDT&E Funding: Base Case Plus T-X, UCLASS, F-22 Foreign Military Sales, and Next-Generation Bomber Programs, FY 2000–2025

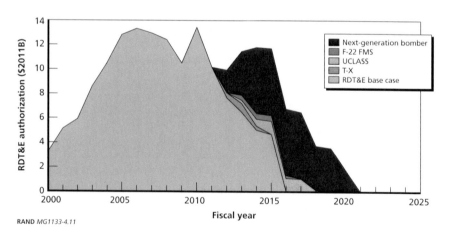

RAND *MG1133-4.11*

Figure 4.12
**Potential Prime Contractor Shares of RDT&E Funding: Base Case Plus
T-X, UCLASS, F-22 Foreign Military Sales, and Next-Generation Bomber
Programs, FY 2010–2025**

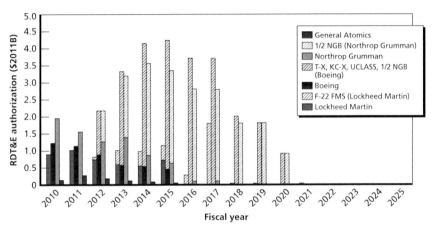

NOTES: NGB = Next-generation bomber.
RAND *MG1133-4.12*

Figure 4.13
**Procurement Funding: Base Case Plus T-X, KC-X, UCLASS, F-22 Foreign
Military Sales, and Next-Generation Bomber Programs, FY 2000–2025**

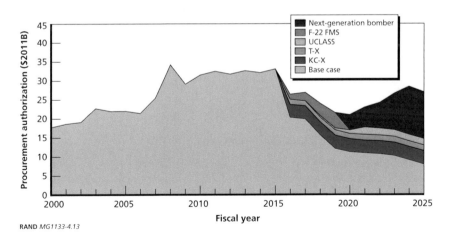

RAND *MG1133-4.13*

Figure 4.14
**Potential Prime Contractor Shares of Procurement Funding: Base Case
Plus T-X, KC-X, UCLASS, F-22 Foreign Military Sales, and Next-Generation
Bomber Programs, FY 2010–2025**

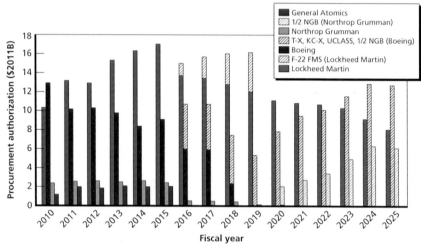

NOTE: NGB = Next-generation bomber.

RAND MG1133-4.14

Implications for RDT&E and Procurement of Adding Six Programs (T-X, KC-X, UCLASS, F-22 Foreign Military Sales, Next-Generation Bomber, and Sixth-Generation Fighter) to Baseline

The final additional program we explore would involve adding a sixth-generation fighter to the previous roster of proposed new programs. As can be seen from Figures 4.15 and 4.16, the sixth-generation fighter has an impact similar to that of the next-generation bomber. Assuming that the program is shared between Lockheed Martin and Northrop Grumman, such a strategy would sustain the RDT&E base for three primes through 2020 and for two through 2025.

As displayed in Figures 4.17 and 4.18, procurement of the sixth-generation fighter does not have much of a near-term impact, but does in the middle of the next decade, with three primes having substantial procurement shares.

Figure 4.15
RDT&E Funding: Base Case Plus T-X, UCLASS, F-22 Foreign Military Sales,
Next-Generation Bomber, and Sixth-Generation Fighter, FY 2000–2025

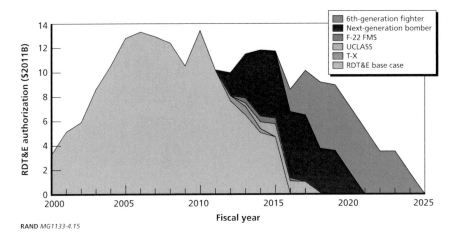

RAND *MG1133-4.15*

Figure 4.16
Potential Prime Contractor Shares of RDT&E Funding: Base Case Plus T-X,
UCLASS, F-22 Foreign Military Sales, Next-Generation Bomber, and
Sixth-Generation Fighter, FY 2010–2025

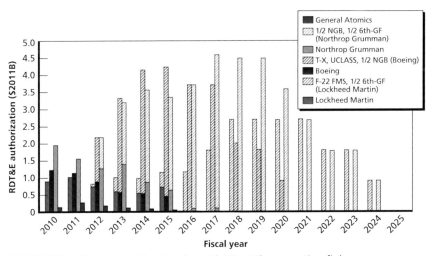

NOTES: NGB = Next-generation bomber. 6th-GF = 6th-generation fighter.
RAND *MG1133-4.16*

Figure 4.17
Procurement Funding: Base Case Plus T-X, KC-X, UCLASS, F-22 Foreign Military Sales, Next-Generation Bomber, and Sixth-Generation Fighter, FY 2000–2025

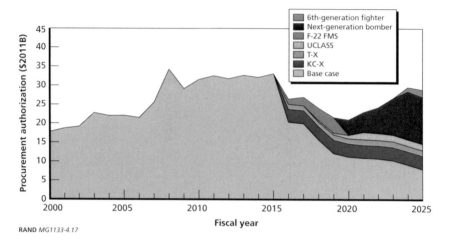

RAND *MG1133-4.17*

Figure 4.18
Potential Prime Contractor Shares of Procurement Funding: Base Case Plus T-X, KC-X, UCLASS, F-22 Foreign Military Sales, Next-Generation Bomber, and Sixth-Generation Fighter, FY 2010–2025

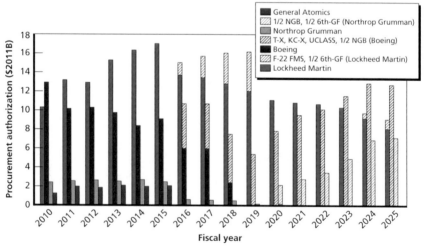

NOTES: NGB = Next-generation bomber. 6th-GF = 6th-generation fighter.
RAND *MG1133-4.18*

Which Program Combinations Would Best Sustain Competition?

It is clear from examining Figures 4.1 through 4.18 that small programs likely will not sustain the industrial base, in terms of either RDT&E or procurement. The T-X, KC-X, and UCLASS programs would, in combination, sustain only Boeing as a viable competitor in the fixed-wing military market, if it were the winner. Selling the F-22 to foreign militaries, as we discuss in Appendix A, would boost Lockheed Martin's fortunes for only four years or so.

What is needed for two primes to remain healthy and competitive in the 2011–2025 period is a new-generation bomber program or a program of similar size. Such a strategy could sustain Boeing and Northrop Grumman if each company were to share 50 percent of the RDT&E funding and procurement funding. After 2025, DoD would need to pursue an additional program on the scale of a sixth-generation fighter. Table 4.2 displays how the primes fare under each of these strategies, with cells in gray and yellow denoting combinations that would significantly or modestly sustain the primes, respectively.

Table 4.2
Program Combinations That Would Sustain Primes in 2011–2025 Period and Post-2025 Period

Time period	Strategy	Boeing	Lockheed Martin	Northrop Grumman
2011–2025	T-X + KC-X + UCLASS	⬛ (gray)		
	+ F-22 foreign military sales	⬛ (gray)	▨ (yellow)	
	+ next-generation bomber	⬛ (gray)	▨ (yellow)	⬛ (gray)
Post-2025	+ sixth-generation fighter	⬛ (gray)	▨ (yellow)	⬛ (gray)

NOTE: Gray = significantly sustained; yellow = modestly sustained.

CHAPTER FIVE

Policy Options Open to the Department of Defense

As noted in Chapter One, we stayed close to the concerns of Congress with regard to maintaining the military aerospace industry's prime contractor competitive structure, and we used Congress' legislative language ("that the United States must ensure, among other things, that more than one aircraft company can design, engineer, produce and support military aircraft in the future") in choosing a criterion to gauge the adequacy of the U.S. military fixed-wing aircraft industrial base.

We interpreted that language to mean that the U.S. industrial base would be adequate if it were able to sustain at least two full-service prime contractors, each possessing approximately equal shares of both RDT&E funding and procurement funding.[1]

Given this interpretation, we found that even though RDT&E funding is at a 30-year high, the industry does not have two primes receiving approximately equal shares of RDT&E funding today. Lockheed Martin, which gets the bulk of funding for the F-35, dominates the picture. Nevertheless, procurement has been more evenly balanced between two of the primes in recent years.

But what if DoD were to pursue additional programs? We identified six new programs—the T-X trainer, the KC-X tanker, the unmanned carrier-launched surveillance and strike aircraft (UCLASS), the F-22 fighter sold as foreign military sales (which we term F-22 FMS), the next-generation bomber, and the sixth-generation fighter—that DoD might consider pursuing, and we mod-

---

[1] See U.S. House of Representatives, 2009, p. 380.

eled the degree to which they might foster innovative, competitive conditions in the future.

Using the 2011 FYDP as our funding base case, we found that if DoD were to award three new programs—T-X, KC-X, and UCLASS—to Boeing (the prime that the research team identified as most in need of additional work), the industry may still become uncompetitive after 2015. Adding the three programs would not change the RDT&E base case picture significantly, nor would it appreciably change the procurement picture.

Adding sales of the F-22 fighter to foreign customers to those three new programs would not improve the industry's competitive picture. Such sales would boost Lockheed Martin's RDT&E fortunes for about four years, and the procurement benefits that it would realize would be concentrated in the 2016–2019 time period.

If, however, DoD were also to involve two primes equally in performing RDT&E and procurement on a next-generation bomber, it could sustain two firms. The bomber would have a significantly larger impact on the RDT&E base case than the four other programs and would sustain two primes through FY 2020. If Boeing and Northrop Grumman were to share the next-generation bomber, they would receive the bulk of RDT&E funding over the next decade. Additionally, such a strategy could reverse a decline in procurement funding and sustain two firms, assuming that Boeing and Northrop Grumman were awarded equal shares. In particular, this would improve Northrop Grumman's procurement outlook.

Adding a sixth-generation fighter to the previous roster of proposed new programs would have an impact similar to that of the next-generation bomber. Assuming that the program would be shared between Lockheed Martin and Northrop Grumman, such a strategy would sustain the RDT&E base for three primes through 2020 and for two primes through 2025. Procurement funding for the sixth-generation fighter would not have much of a near-term impact, but it would be more influential in the middle of the next decade, with three primes having substantial procurement shares.

But all these projections of the effect of RDT&E and procurement funding need to be viewed in light of the changing business models and innovation sources that have emerged over the past decade. The interplay between these new, evolving business practices and innovation and competition has not been examined extensively and is an area for further study. Nonetheless, there is little doubt that primes are looking to lower-level partner firms as promising seedbeds of cutting-edge development, innovation, and discovery. Whereas primes were the predominant sources of competitive developments in the past, today lower-level firms are also in the competitive mix.

The aerospace industry appears to be morphing toward commercial enterprise models that rely on networks of agile, smaller teams that have autonomy, budgets, and delayered authority structures and processes. Primes now outsource much of what they once did in house.

As a single buyer, DoD is in a unique position to foster innovation in the defense industry. But to do that, DoD needs to better understand capital markets and to develop policies to interact more effectively with small and medium-sized companies to improve innovation.

Our findings indicate that procurement funding likely will be adequate to sustain the basic institutional structure of the current prime military aircraft contractors through at least the end of the present decade. New R&D activities with a high likelihood of occurrence (a new trainer, a new tanker, UCLASS) may be sufficient to sustain the design and development capabilities of the current primes through the middle of this decade. However, aircraft such as tankers that would be derived from commercial platforms and UAS/UAV/UCAV programs as currently planned will be insufficient to sustain the current industry structure and capabilities beyond this decade. A DoD decision to begin a new major combat aircraft program before the end of this decade would provide a stronger basis for sustaining the current structure and capability. Conversely, if the number and frequency of major aircraft programs continue to diminish, it will be increasingly difficult to sustain a competition-based industry of the size and posture that exists today.

A variety of new factors since 2000 suggests that it may be possible to maintain a competitive and innovative fixed-wing military aircraft industrial base into the immediate future, even with a reduced number of prime contractors and new program starts during a period of growing pressure on the federal budget. This is mainly because of three new trends that have emerged over the last ten years: the dramatic upsurge in RDT&E and procurement funding following the attacks of September 11, 2001; the large increase in the development and procurement of UAS, accompanied by the entry of new contractors and regeneration of traditional firms; and the continuing movement toward greater competitive outsourcing of research, development, and production tasks to lower-tier contractors, both foreign and domestic.

In this study, we have argued that the future composition and capabilities of the military aircraft industry depend largely on the amount of business that the industry receives from DoD and how that business is distributed among development of technology, development of new designs, and production of completed designs. In Chapter Four, we assumed that those firms that most needed the work won the award. However, competition may not produce the outcomes discussed, and the industry may concentrate further. Directed shares may be necessary to sustain multiple primes into the foreseeable future. Unless very purposeful and structured program decisions are made soon, the congressional objective—that the United States maintain two or more companies capable of designing, engineering, producing, and supporting military aircraft—will not be achieved.

F-22 Foreign Military Sales: Implications for the U.S. Fixed-Wing Military Aircraft Industrial Base

Sections 1250(a) and 1250(b) of the 2010 National Defense Authorization Act (NDAA) required a report on the cost, technical feasibility, strategic implications, and legal changes required for the development of an export variant of the F-22A fighter. That report was submitted in early May 2010. Section 1250(c) of the NDAA required an FFRDC to prepare an additional report

> . . . [o]n the impact of foreign military sales of the F-22A fighter aircraft on the United States aerospace and aviation industry, and the advantages and disadvantages of such sales for sustaining that industry.

This appendix meets the requirements of Section 1250(c). It analyzes the impact on the U.S. military aircraft industrial base of exporting an FMS version of the U.S. F-22A fighter aircraft. To prepare it, we took the following approach:

- We consolidated existing unclassified research on the F-22A supply chain and F-22 FMS variants. Although we did not identify specific components requiring additional nonrecurring effort to make an FMS variant and the firms that supply them, we provided a broad understanding of modifications required.
- We examined the type, scope, and cost of modifications required by previous fighter FMS programs.

- We compared the cost of the modifications to the rest of the aircraft R&D.
- We examined the benefit of F-22 FMS R&D and procurement funding for the overall fixed-wing military aircraft industrial base.
- We noted that the likely substitute for an F-22 FMS variant would be other U.S. produced-fighter aircraft, mitigating the benefits of F-22 FMS sales for the industrial base.

Outline of the F-22A Industrial Base

The industrial base supporting for the F-22A fighter cuts across a large portion of the fixed-wing military aircraft industry. Lockheed Martin (Marietta, Georgia, and Fort Worth, Texas) and Boeing (Puget Sound, Washington) are the prime air vehicle contractors for the F-22A; Pratt & Whitney (East Hartford and Middletown, Connecticut) is the prime engine contractor. Hundreds of suppliers contribute mission and vehicle systems, including Northrop Grumman, Raytheon, Honeywell, and BAE. Titanium is sourced from Timet. Composite raw materials are sourced from Cytec. Major structures are produced by Wyman Gordon, GKN, and many others. A number of firms use specialized facilities, capital equipment, government tooling, and skilled labor to make these parts. Because the major subsystems are so complex, they are produced on a sole-source basis.

Current Status of the F-22A Industrial Base

At the time that this appendix was written in 2010, the last four F-22A fighter aircraft were purchased by the Air Force. Because an F-22 FMS variant was not in the planning horizon for the F-22 team, production shutdown activities have started on a site-by-site basis as the last F-22A components are being produced. Several lowest-tier suppliers have already delivered their last units, a few higher-tier suppliers will be delivering their last units imminently, and one prime contractor facility (Lockheed Martin, Fort Worth) has already shrunk its footprint to accommodate other aircraft in production. However, the change is not irreversible: Major capital equipment has not yet been sold or transferred, and government tooling is being preserved.

Effects of F-22 FMS Restart on the Industrial Base

Any procurement of an F-22 FMS variant, even beginning early in FY 2011, would represent an unanticipated and dramatic shift in plans by the Air Force and contractor teams. By the time F-22 FMS production began, the F-22 production line would have been shut down for five years. Hence, an F-22 FMS variant would require an extensive and expensive production restart program at many component, subsystem, and prime facilities.

Concurrent with these nonrecurring restart activities, the F-22 FMS would have an engineering and manufacturing development (EMD) period of up to four years. Consistent with previous studies of F-22A restart, production would then ramp up to from five units in FY 2015 to 10 in FY 2016 to 20 units in FY 2017 through FY 2019.

Effects on the Overall Industrial Base

The effect of F-22 FMS RDT&E and procurement funding on the overall industrial base is uncertain. As we discussed in relation to Figures 4.15 and 4.17, F-22 FMS would have a modest impact on the RDT&E and procurement pictures, and would last only four years. It would largely affect only one prime, Lockheed Martin.

Protecting Sensitive or Critical Technologies

The United States has substantial historical and current experience in FMS of combat aircraft. The late 1990s and early 2000s have seen the export of FMS variants of the Air Force's F-15 and F-16, as well as the Navy's F/A-18, and AV-8B. The F-35 will have versions for international partners and an FMS version for non-partner nations. The processes and procedures for initiating, marketing, and development are standard.

U.S. law prohibits the export of critical technologies. Hence, specific nonrecurring effort would have to be taken to modify hardware and software from that of the F-22A to an F-22 FMS version. The cost of such efforts is borne by the customer, but it is not work that sustains or enhances the industrial base to a considerable degree. In light of the classification of those technologies, we will only outline the technology protection efforts required.

Lockheed Martin has claimed that it could have developed an F-22 FMS variant that has highly common systems and software with the F-22A and that its anti-tamper technology would have involved existing designs, hardware, and software already assessed and approved to protect similar critical technologies.

How the F-22A Industrial Base Is Unique

Several published reports highlight unique or critical worker skills, knowledge, and processes used in making parts for the F-22A. Previous RAND research has documented these and other issues in the F-22A supply base. But that research focused on unique components and systems that would face disparate impacts either on production shutdown or on restart. In those studies, the suppliers with multiple lines of business producing similar systems for other aircraft were of less concern that those with the unique industrial capabilities required to build F-22A aircraft. Because F-22A production is shutting down, those concerns persist for any F-22 FMS version.

F-22A Diminishing Manufacturing Sources

Even when in low-rate production, the F-22A program has continually faced suppliers who found that it was no longer in their best interest to supply critical, unique electronics components to the prime contractors. Multiple strategies have been implemented to counter this difficulty, from finding and setting up alternative suppliers to outright pre-purchase of parts for future aircraft. While a modernization program will continue after shutdown, keeping many suppliers in a low-rate delivery status, diminishing manufacturing sources (DMS) will still be an issue for some electronics components. The addition of an F-22 FMS variant is likely to slow, but not substantially alter, DMS on the F-22 program.

Meeting Requirements Without an F-22 FMS Variant

As stated above, the production of an F-22 FMS variant would have an effect, though not significant, on the overall industrial base. Billions of dollars in R&D and procurement funding would enrich or sustain the business of several hundred firms and require continued employment of thousands of full-time-equivalent workers in the United States for several years. Perhaps most importantly, particular F-22 unique skills and production process would be maintained.

However, these effects must be considered in light of the alternative ways that FMS customers might meet their aircraft requirements. If an F-22 FMS variant is not produced, foreign nations that have expressed an interest in it will likely satisfy their requirements with other fourth- or fifth-generation fighters. An interest in F-22 capabilities indicates a requirement for stealth and air superiority, which could be filled by a high-low mix of globally sourced aircraft. As stated above, U.S. firms have three fourth-generation airframes (with fifth-generation avionics) already on the export market and available for delivery within a few years. An F-35 export variant will not be available for several years.

The export of any fighter aircraft by a U.S. firm is likely to have similar high-level effects on the U.S. industrial base, but the details of each program mean that different capabilities would be maintained. Substituting FMS of another aircraft for F-22 FMS would have four broad effects.

1. Existing FMS versions already exist for other fighter aircraft, while F-22 will require an EMD program. This might transfer to a smaller overall level of funding, or might turn into a larger number of aircraft, or more advanced versions of aircraft, procured.

2. Suppliers of components unique to the non-export F-22 would not be sustained.

3. Production of F-22 airframe structures, engines, vehicle systems, and mission systems is primarily done in the United States. Other aircraft have large components, structures, and entire subsystems manufactured globally. For example, KAI of South Korea manufactures forward fuselages and wings for the F-15, and TAI of Turkey manufactures the all-composite air inlet ducts for the F-35.[1]

4. Different prime contractors would benefit, changing the pattern of employment. While the F-22 is a team effort between Lockheed Martin and Boeing, other fighters are produced by Boeing (F-15), Boeing and Northrop Grumman (F/A-18), Lockheed Martin (F-16), or Lockheed Martin and Northrop Grumman (F-35). Each of those has an existing supply chain. Estimating these effects would require a detailed supply chain analysis by major element.

Alternatively, if those customers procured an aircraft from a non-U.S. firm, there would be some, but substantially less, effect on the U.S. industrial base.

---

[1] "TAI Delivers First Large F-35 Composite Structure," Netcomposites.com, August 13, 2010.

RDT&E and Procurement: RAND 2003 Funding Projections Compared with Actual Funding, FY 2003–2010

In this appendix, we compare the funding projections that RAND made in its 2003 congressionally mandated study of the fixed-wing aircraft military industrial base[1] with actual obligational authority for RDT&E and procurement that DoD received from FY 2000 through FY 2010 and with funding that the FY 2011 FYDP projects DoD will receive over the next 10–15 years.

To accurately portray these data, we first reproduce graphs showing the RDT&E and procurement predictions that RAND published in its 2003 study. We then show those graphs translated into FY 2011 dollars. Finally, we overlay on those predictions a line that shows (a) actual funding that DoD received from FY 2000 through FY 2010 and (b) funding that DoD is expected to receive from FY 2011 through FY 2020 or FY 2025 as outlined in the FY 2011 FYDP. For clarity, the portion of the line that shows actual funding is solid, and the portion that shows the FY 2011 FYDP data is dashed.

The base case that we depicted in 2003 showed TOA, which included funding that went to prime contractors, funding for contractors that provided Government Furnished Equipment (GFE) to the primes, and funding for government organizations' costs directly related to the program. In our calculations of RDT&E activities, the prime contractors received 70 to 90 percent of TOA.

The additional programs that we postulated in 2003 included programs that we believed were highly likely to start in the first few

---

[1] Birkler et al., 2003.

years of the decade and more speculative ones that started later in the decade. Each program represented different dollar levels of activity and different kinds of design and development work and allowed us to get some idea of how those two parameters might interact to support the industry in different periods. The near-term programs that we postulated were wide-body transport derivatives that would replace aging aircraft in several roles (tankers and multimission command-and-control aircraft, which we showed as ISR in our graphics) and a UCAV weapon system. The additional, more speculative program that we postulated in 2003 was a new major combat aircraft (MCA).

Figures B.1 through B.3 show RAND's predictions, actual outlays, and FY 2011 FYDP planned outlays for RDT&E funding. Figures B.4 thorough B.6 show similar displays for procurement funding.

In terms of RDT&E spending, actual and FY 2011 FYDP planned outlays exceed the levels that we predicted in 2003 until FY 2016, after which the planned spending falls below our prediction.

In terms of procurement outlays, actual spending tracked our 2003 prediction from FY 2000 through FY 2007, after which it

Figure B.1
2003 RDT&E Prediction, Reproduction of Figure 4.13 in RAND MR-1656-OSD: RDT&E Obligational Authority, Base Case Plus Postulated Near-Term Programs (UCAV, ISR, and Tanker), and Major Combat Aircraft (MCA), FY 2003 Dollars

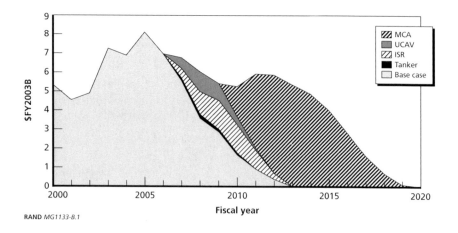

Figure B.2
2003 RDT&E Prediction, FY 2011 Dollars

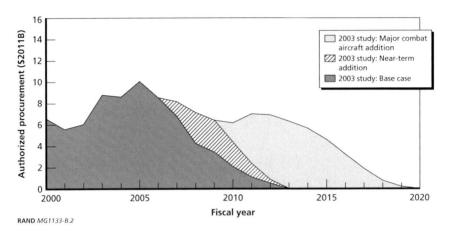

RAND *MG1133-B.2*

Figure B.3
2003 RDT&E Prediction Compared with FY 2000–2010 Actual Funding and FY 2011 FYDP Planned Funding, FY 2011 Dollars

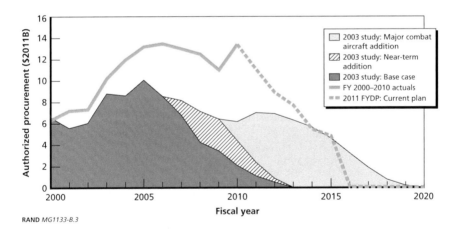

RAND *MG1133-B.3*

exceeded our predicted level through FY 2010. Between FY 2011 and FY 2014, the spending in the FYDP plan exceeds our predicted level. After then, however, the spending outlined in the FY 2011 FYDP is

Figure B.4
2003 Procurement Prediction, Reproduction of Figure 4.14 in RAND MR-1656-OSD: Procurement Obligational Authority, Base Case Plus Postulated Near-Term Programs (UCAV, ISR, and Tanker), and Major Combat Aircraft (MCA), FY 2003 Dollars

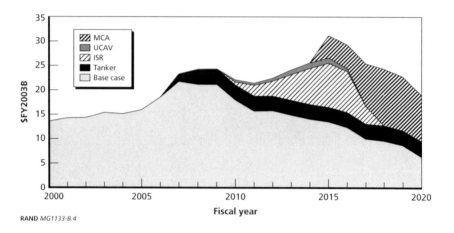

RAND *MG1133-B.4*

Figure B.5
2003 Procurement Prediction, FY 2011 Dollars

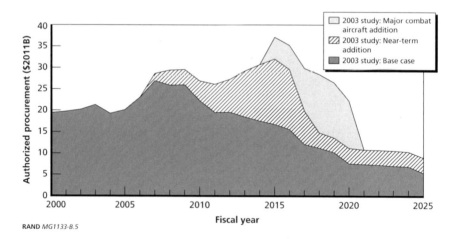

RAND *MG1133-B.5*

Figure B.6
**2003 Procurement Prediction Compared with FY 2000–2010 Actual Funding
and FY 2011 FYDP Planned Funding, FY 2011 Dollars**

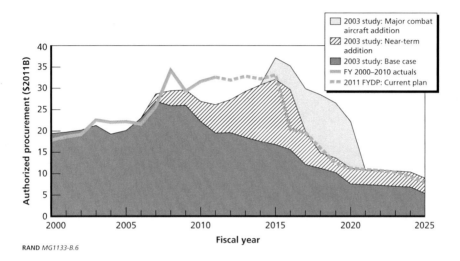

either less than the amount we predicted (from FY 2015 through
FY 2021) or equal to our earlier prediction (from FY 2022 through
FY 2025).

RDT&E and Procurement Funding: Contractor Shares in Program of Record and Projected Outlays for New Programs in the FY 2011 FYDP

This appendix gives readers a different view of the data that we discussed in Chapter Four. It compares the amounts of RDT&E and procurement funding that Lockheed Martin, Boeing, Northrop Grumman, and General Atomics are likely to receive from fixed-wing military aircraft projects in the program of record with total RDT&E and procurement outlays that the six new programs explored in Chapter Four are projected to require.

To portray these data, we reproduce elements of the figures in Chapter Four that show the shares of RDT&E and procurement funding going to prime contractors in the program of record base case. We overlay, on top of those displays, lines that show the total amount of RDT&E and procurement funding associated with the six programs in the four scenarios that we explored in Chapter Four. As a reminder, those programs and scenarios were as follows:

1. Acquisition (RDT&E plus procurement) of three new programs: the T-X trainer, the KC-X tanker, and the UCLASS
2. Acquisition of four new programs: Scenario 1 plus the F-22 fighter sold as foreign military sales (F-22 FMS)
3. Acquisition of five new programs: Scenario 2, plus the next-generation bomber
4. Acquisition of six new programs: Scenario 3, plus the sixth-generation fighter.

The analysis in Chapter Four apportioned the funding connected with the *new* programs among various prime contractors. This appendix does not apportion that new program funding. Instead, in Figures C.1–C.10, it shows funding in terms of total projected outlays.

Figure C.1
RDT&E Funding: Contractor Shares in Program of Record and Projected Total Outlays in FY 2011 FYDP for T-X and UCLASS Programs, FY 2010–2025

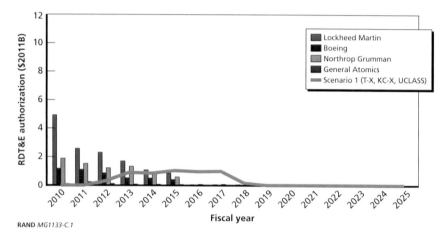

RAND *MG1133-C.1*

Figure C.2
Procurement Funding: Contractor Shares in Program of Record and Projected Total Outlays in FY 2011 FYDP for T-X, KC-X, and UCLASS Programs, FY 2010–2025

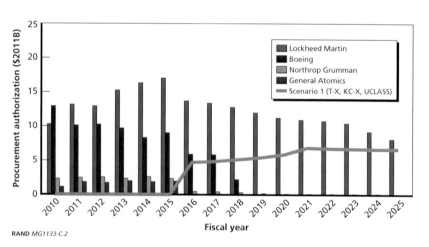

RAND *MG1133-C.2*

Figure C.3
RDT&E Funding: Contractor Shares in Program of Record and Projected
Total Outlays in FY 2011 FYDP for T-X, UCLASS, and F-22 Foreign Military
Sales Programs, FY 2010–2025

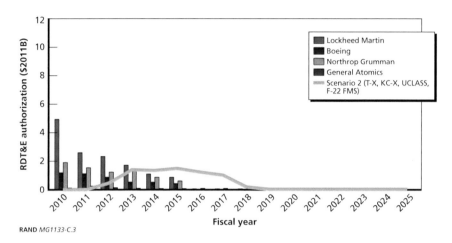

RAND MG1133-C.3

Figure C.4
Procurement Funding: Contractor Shares in Program of Record and
Projected Total Outlays in FY 2011 FYDP for T-X, KC-X, UCLASS, and F-22
Foreign Military Sales Programs, FY 2010–2025

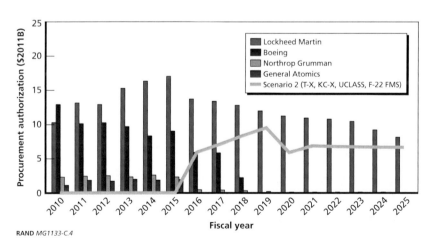

RAND MG1133-C.4

Figure C.5
RDT&E Funding: Contractor Shares in Program of Record and Projected
Total Outlays in FY 2011 FYDP for T-X, UCLASS, F-22 Foreign Military Sales,
and Next-Generation Bomber Programs, FY 2010–2025

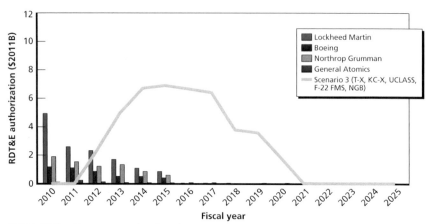

NOTE: NGB = Next-generation bomber.
RAND MG1133-C.5

Figure C.6
Procurement Funding: Contractor Shares in Program of Record and
Projected Total Outlays in FY 2011 FYDP for T-X, KC-X, UCLASS, F-22
Foreign Military Sales, and Next-Generation Bomber Programs,
FY 2010–2025

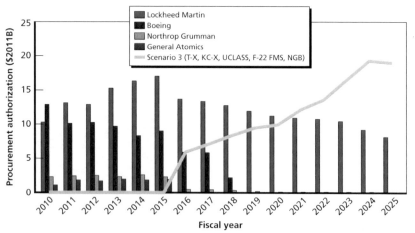

NOTE: NGB = Next-generation bomber.
RAND MG1133-C.6

Figure C.7
RDT&E Funding: Contractor Shares in Program of Record and Projected Total Outlays in FY 2011 FYDP for T-X, UCLASS, F-22 Foreign Military Sales, Next-Generation Bomber, and Sixth-Generation Fighter, FY 2010–2025

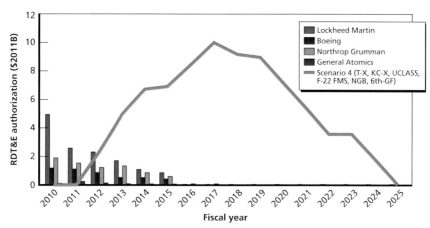

NOTES: NGB = Next-generation bomber. 6th-GF = 6th-generation fighter.
RAND *MG1133-C.7*

Figure C.8
Procurement Funding: Contractor Shares in Program of Record and Projected Total Outlays in FY 2011 FYDP for T-X, KC-X, UCLASS, F-22 Foreign Military Sales, Next-Generation Bomber, and Sixth-Generation Fighter, FY 2010–2025

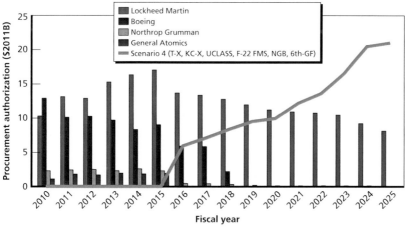

NOTES: NGB = Next-generation bomber. 6th-GF = 6th-generation fighter.
RAND *MG1133-C.8*

Figure C.9
RDT&E Funding: Contractor Shares in Program of Record and Projected Total Outlays in FY 2011 FYDP for All New Program Scenarios FY 2010–2025

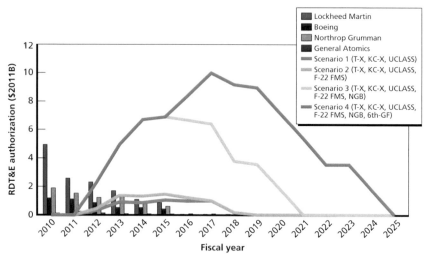

NOTES: NGB = Next-generation bomber. 6th-GF = 6th-generation fighter.
RAND MG1133-C.9

Figure C.10
Procurement Funding: Contractor Shares in Program of Record and
Projected Total Outlays in FY 2011 FYDP for All New Program Scenarios,
FY 2010–2025

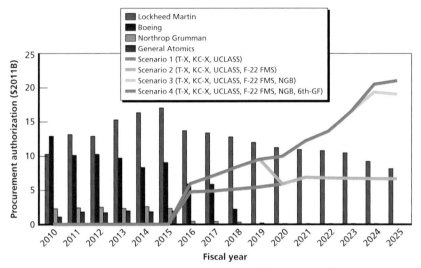

NOTES: NGB = Next-generation bomber. 6th-GF = 6th-generation fighter.

RAND MG1133-C.10

U.S. Total Military Air Vehicle Procurement Quantities, FY 2012–2021

This appendix provides readers with a snapshot of the quantities of air vehicles that the U.S. military services will acquire with the RDT&E and procurement funding that we discuss in Chapter Four. Note that while the program-of-record figures in Chapter Four show funding through FY 2025, Table D.1 and Figure D.1 (on the following pages) show only inventories through FY 2021.

Table D.1
Aviation Investment Plan Taxonomy

	Fighter/Attack	Unmanned Multirole Surveillance and Light Strike	ISR/C2	Intratheater Lift	Strategic Lift	Aerial-Refueling Tanker	Bomber
U.S. Air Force	A-10 F-15C/D F-15E F-16 F-22 F-35A	MQ-9 F/O UAS[a]	MQ-1, RQ-4, U-2, E-3, E-4, E-8, EC-130, EC-130 Recap, RC-26, RC-135, MC-12, WC-135	C-130E/H C-130J C-27 WC-130J/H	C-5 C-17	KC-10 KC-46A KC-135	B-1 B-2 B-52 LRS-B[b]
DoN	AV-8B EA-6B EA-18G F/A-18A-D F/A-18E/F F-35B, F-35C NGAD[c]	UCLASS USMC Grp 4[d] F/O UAS	MQ-4, E-2C E-2D, E-6. P-3 EP-3, P-8	AR/LSB[e] C-130J, C-2 C-9, C-12 C-12 Recap C-20, C-26 C-37, C-40		KC-130J KC-130T	

NOTE: DoN = Department of the Navy.
[a] Follow-on UAS.
[b] Long-Range Strike Bomber.
[c] Next Generation Air Dominance.
[d] U.S. Marine Corps Group 4 Unmanned Aerial System.
[e] Airborne Resupply/Logistics for the Sea-Base.

Figure D.1
Total U.S. Military Aircraft Procurement Quantities, FY 2012–2021

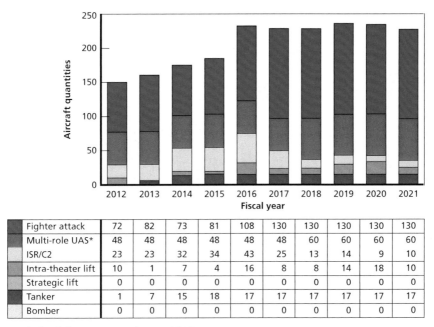

	2012	2013	2014	2015	2016	2017	2018	2019	2020	2021
Fighter attack	72	82	73	81	108	130	130	130	130	130
Multi-role UAS*	48	48	48	48	48	48	60	60	60	60
ISR/C2	23	23	32	34	43	25	13	14	9	10
Intra-theater lift	10	1	7	4	16	8	8	14	18	10
Strategic lift	0	0	0	0	0	0	0	0	0	0
Tanker	1	7	15	18	17	17	17	17	17	17
Bomber	0	0	0	0	0	0	0	0	0	0

*Includes follow-on UAS after FY 2016.

RAND *MG1133-D.1*

Bibliography

Aerospace Industries Association, *Launching the 21st Century American Aerospace Workforce,* Arlington, Va., December 2008. As of July 17, 2011:
http://www.aia-aerospace.org/pdf/report_workforce_1208.pdf

Airbus SAS, "Orders & Deliveries. The Month in Review: June 2011." As of July 17, 2011:
http://www.airbus.com/company/market/orders-deliveries/

American Association of Engineering Societies, Engineering Workforce Commission, *Engineering & Technology Enrollments*, various years. As of July 19, 2011:
http://www.ewc-online.org/data/enrollments_data.asp

American Institute of Aeronautics and Astronautics (AIAA), "Recruiting, Retaining, and Developing a World-Class Aerospace Workforce: An AIAA Information Paper." As of August 17, 2010:
http://pdf.aiaa.org/downloads/publicpolicypositionpapers/Retaining%20Aero%20Workforce.pdf

"Aviation Week 2009 Workforce Study," *Aviation Week*, August 24, 2009.

Birkler, John, Mark V. Arena, Irv Blickstein, Jeffrey A. Drezner, Susan M. Gates, Meilinda Huang, Robert Murphy, Charles Nemfakos, and Susan K. Woodward, *From Marginal Adjustments to Meaningful Change: Rethinking Weapon System Acquisition*, Santa Monica, Calif.: RAND Corporation, MG-1020-OSD, 2010. As of July 17, 2011:
http://www.rand.org/pubs/monographs/MG1020.html

Birkler, John, Anthony G. Bower, Jeffrey A. Drezner, Gordon Lee, Mark Lorell, Giles Smith, Fred Timson, William P.G. Trimble, and Obaid Younossi, *Competition and Innovation in the U.S. Fixed-Wing Military Aircraft Industry*, Santa Monica, Calif.: RAND Corporation, MR-1656-OSD, 2003. As of July 17, 2011:
http://www.rand.org/pubs/monograph_reports/MR1656.html

Boeing Company, Orders and Deliveries. July 12, 2011. As of July 20, 2011:
http://active.boeing.com/commercial/orders/index.cfm

Bumiller, Elisabeth, and Michael Wines, "Test of Stealth Fighter Clouds Gates Visit to China," *New York Times*, January 11, 2011. As of July 20, 2011: http://www.nytimes.com/2011/01/12/world/asia/12fighter.html

Datamonitor USA, *Global Aerospace & Defense Industry Profile*, New York: Datamonitor Publication 0199-1002, December 2006.

Defense Acquisition University, ACQuipedia, Program Objective Memorandum (POM/Budget Submit [on-year]), May 2010. As of July 21, 2011: https://acc.dau.mil/ILC_POMBSOY

Department of Defense, *Aircraft Procurement Plan Fiscal Years (FY) 2012–2041: Submitted with the FY 2012 Budget*, Washington, D.C.: Department of Defense, RefID A-A78844C, March 11, 2011.

Eisenmann, Thomas R., "Managing Networked Businesses: Course Overview for Educators," Harvard Business School Case 5-807-104, Revised January 2, 2007. As of July 20, 2011: http://www.hbs.edu/units/em/pdf/EisenmannMNBoverviewTN-for-PDF_3-8.pdf

Foster, Richard N., *Innovation: The Attacker's Advantage*, New York: Summit Books, 1986.

Gansler, Jacques S., *The Defense Industry*, Cambridge, Mass.: MIT Press, 1982.

Gassman, Oliver, Gerrit Reepmeyer, and Maxmillian von Zedtwitz, *Leading Pharmaceutical Innovation: Trends and Drivers for Growth in the Pharmaceutical Industry*, Berlin: Springer, 2008.

Gulati, Ranjay, *Managing Network Resources, Alliances, Affiliations, and Other Relational Assets*, Oxford, UK: Oxford University Press, 2006.

———, "Silo Busting, How to Execute on the Promise of Customer Focus," *Harvard Business Review*, Vol. 12, No. 5, May 2007.

———, *Reorganize for Resilience, Putting Customers at the Center of Your Business*, Boston, Mass.: Harvard Business Press, 2009.

Holstein, William J., *The Next American Economy*, New York: Walker & Co., 2011.

Johnson, Mark W., *Seizing the White Space: Business Model Innovation for Growth and Renewal*, Boston, Mass.: Harvard Business Press, 2010.

Leifer, Richard, et al., *Radical Innovation: How Mature Companies Can Outsmart Upstarts*, Boston, Mass.: Harvard Business School Press, 2000.

Lorell, Mark, *The U.S. Combat Aircraft Industry, 1909–2000*, Santa Monica, Calif.: RAND Corporation, MR-1696-OSD, 2003. As of July 20, 2011: http://www.rand.org/pubs/monograph_reports/MR1696.html

Martins, CAPT John, *Joint Strike Fighter Program Update*, slide presentation, n.d. As of July 19, 2011: http://www.dtic.mil/ndia/2009psa_mar/Martins.pdf

McKern, Bruce, ed., *Managing the Global Network Corporation*, London: Routledge, 2003.

Ministry of Defence, Government of India, *Defence Procurement Procedure 2008*, New Delhi, July 2008. As of July 20, 2011:
http://mod.nic.in/dpm/welcome.html

Nanto, Dick K., *Globalized Supply Chains and U.S. Policy*, Washington, D.C.: Congressional Research Service, CRS Report R40167, January 27, 2010.

National Defense Authorization Act for Fiscal Year 2010, Public Law 111-84, October 28, 2009, Section 1250.

National Science Board, *Science and Engineering Indicators 2010*, Arlington, Va.: National Science Foundation, 2010. As of July 19, 2011:
http://www.nsf.gov/statistics/seind10/pdf/seind10.pdf

Nelson, Richard, ed., *National Innovation Systems*, Oxford, UK: Oxford University Press, 1993.

Peck, Merton J., and Frederic M. Scherer, *The Weapons Acquisition Process: An Economic Analysis*, Boston, Mass.: Harvard Business School Division of Research, 1962.

Perrow, Charles, *Complex Organizations*, New York: Random House, 1986.

Platzer, Michaela D., *U.S. Aerospace Manufacturing: Industry Overview and Prospects*, Washington, D.C.: Congressional Research Service, CRS Report R40967, December 3, 2009.

Podolny, Joel, "Social Capital," briefing slides, New Haven, Conn.: Yale School of Management, 2007.

Roberts, John, *Organizational Design for Performance and Growth*, Oxford, UK: Oxford University Press, 2006.

Skarzynski, Peter, and Rowan Gibson, *Innovation to the Core: A Blueprint for Transforming the Way Your Company Innovates*, Boston, Mass.: Harvard Business Press, 2008.

———, *Building a Systematic Innovation a Capability*, Boston, Mass.: Harvard Business Press, 2009.

Spatareanu, Mariana, "The Cost of Capital, Finance and High-Tech Investment," *International Review of Applied Economics*, Vol. 22, No. 6, November 2008.

"TAI Delivers First Large F-35 Composite Structure," Netcomposites.com, August 13, 2010. As of July 21, 2011:
http://www.netcomposites.com/newspic.asp?6239

Thomke, Stefan, and Barbara Feinberg, *Design Thinking and Innovation at Apple*, Harvard Business School Case 9-609-066, March 4, 2010.

U.S. Census Bureau: Maufacturers and Aircraft Manufacturing, various years.

U.S. Department of Defense, Office of the Under Secretary of Defense for Acquisition, Technology, and Logistics Industrial Policy, *Annual Industrial Capabilities Report to Congress*, March 2009. As of July 19, 2011:
http://www.acq.osd.mil/mibp/docs/annual_ind_cap_rpt_to_congress-2009.pdf

U.S. Department of Justice, "The Herfindahl-Hirschman Index," n.d. As of July 19, 2011:
http://www.justice.gov/atr/public/testimony/hhi.htm

U.S. House of Representatives, National Defense Authorization Act of 2010: Report of the Committee on Armed Services, House of Representatives, on H.R. 2647, Together with Additional and Supplemental Views, Report 111–166, Washington, D.C., U.S. Government Printing Office, June 18, 2009. As of July 20, 2011:
http://www.gpo.gov/fdsys/pkg/CRPT-111hrpt166/html/CRPT-111hrpt166.htm

Younossi, Obaid, Kevin Brancato, John C. Graser, Thomas Light, Rena Rudavsky, and Jerry M. Sollinger, *Ending F-22A Production: Costs and Industrial Base Implications of Alternative Options*, Santa Monica, Calif.: RAND Corporation, MG-797-AF, 2010. As of July 19, 2011:
http://www.rand.org/pubs/monographs/MG797.html